Waste Stabilisation Ponds

Biological Wastewater Treatment Series

The *Biological Wastewater Treatment* series is based on the book *Biological Wastewater Treatment in Warm Climate Regions* and on a highly acclaimed set of best selling textbooks. This international version is comprised by six textbooks giving a state-of-the-art presentation of the science and technology of biological wastewater treatment.

Titles in the *Biological Wastewater Treatment* series are:

Volume 1: *Wastewater Characteristics, Treatment and Disposal*
Volume 2: *Basic Principles of Wastewater Treatment*
Volume 3: *Waste Stabilisation Ponds*
Volume 4: *Anaerobic Reactors*
Volume 5: *Activated Sludge and Aerobic Biofilm Reactors*
Volume 6: *Sludge Treatment and Disposal*

Biological Wastewater Treatment Series

VOLUME THREE

Waste Stabilisation Ponds

Marcos von Sperling

Department of Sanitary and Environmental Engineering
Federal University of Minas Gerais, Brazil

Published by IWA Publishing, Alliance House, 12 Caxton Street, London SW1H 0QS, UK

Telephone: +44 (0) 20 7654 5500; Fax: +44 (0) 20 7654 5555; Email: publications@iwap.co.uk
Website: **www.iwapublishing.com**

First published 2007
© 2007 IWA Publishing

Copy-edited and typeset by Aptara Inc., New Delhi, India

British Library Cataloguing in Publication Data
A CIP catalogue record for this book is available from the British Library

Library of Congress Cataloguing-in-Publication Data
A catalogue record for this book is available from the Library of Congress

ISBN: 1 84339 163 5
ISBN 13: 9781843391630

Contents

Contents

Preface

The present series of books has been produced based on the book *"Biological wastewater treatment in warm climate regions"*, written by the same authors and also published by IWA Publishing. The main idea behind this series is the subdivision of the original book into smaller books, which could be more easily purchased and used.

The implementation of wastewater treatment plants has been so far a challenge for most countries. Economical resources, political will, institutional strength and cultural background are important elements defining the trajectory of pollution control in many countries. Technological aspects are sometimes mentioned as being one of the reasons hindering further developments. However, as shown in this series of books, the vast array of available processes for the treatment of wastewater should be seen as an incentive, allowing the selection of the most appropriate solution in technical and economical terms for each community or catchment area. For almost all combinations of requirements in terms of effluent quality, land availability, construction and running costs, mechanisation level and operational simplicity there will be one or more suitable treatment processes.

Biological wastewater treatment is very much influenced by climate. Temperature plays a decisive role in some treatment processes, especially the natural-based and non-mechanised ones. Warm temperatures decrease land requirements, enhance conversion processes, increase removal efficiencies and make the utilisation of some treatment processes feasible. Some treatment processes, such as anaerobic reactors, may be utilised for diluted wastewater, such as domestic sewage, only in warm climate areas. Other processes, such as stabilisation ponds, may be applied in lower temperature regions, but occupying much larger areas and being subjected to a decrease in performance during winter. Other processes, such as activated sludge and aerobic biofilm reactors, are less dependent on temperature, as a result of the higher technological input and mechanisation level. The main purpose of this series of books is to present the technologies for urban wastewater treatment as applied to the specific condition of warm temperature, with the related implications in terms of design and operation. There is no strict definition for the range of temperatures that fall into this category, since the books always present how to correct parameters, rates and coefficients for different temperatures. In this sense, subtropical and even temperate climate are also indirectly covered, although most of the focus lies on the tropical climate.

Another important point is that most warm climate regions are situated in developing countries. Therefore, the books cast a special view on the reality of these countries, in which simple, economical and sustainable solutions are strongly demanded. All technologies presented in the books may be applied in developing countries, but of course they imply different requirements in terms of energy, equipment and operational skills. Whenever possible, simple solutions, approaches and technologies are presented and recommended.

Considering the difficulty in covering all different alternatives for wastewater collection, the books concentrate on off-site solutions, implying collection and transportation of the wastewater to treatment plants. No off-site solutions, such as latrines and septic tanks are analysed. Also, stronger focus is given to separate sewerage systems, although the basic concepts are still applicable to combined and mixed systems, especially under dry weather conditions. Furthermore, emphasis is given to urban wastewater, that is, mainly domestic sewage plus some additional small contribution from non-domestic sources, such as industries. Hence, the books are not directed specifically to industrial wastewater treatment, given the specificities of this type of effluent. Another specific view of the books is that they detail biological treatment processes. No physical-chemical wastewater treatment processes are covered, although some physical operations, such as sedimentation and aeration, are dealt with since they are an integral part of some biological treatment processes.

The books' proposal is to present in a balanced way theory and practice of wastewater treatment, so that a conscious selection, design and operation of the wastewater treatment process may be practised. Theory is considered essential for the understanding of the working principles of wastewater treatment. Practice is associated to the direct application of the concepts for conception, design and operation. In order to ensure the practical and didactic view of the series, 371 illustrations, 322 summary tables and 117 examples are included. All major wastewater treatment processes are covered by full and interlinked design examples which are built up throughout the series and the books, from the determination of the wastewater characteristics, the impact of the discharge into rivers and lakes, the design of several wastewater treatment processes and the design of the sludge treatment and disposal units.

The series is comprised by the following books, namely: *(1) Wastewater characteristics, treatment and disposal; (2) Basic principles of wastewater treatment; (3) Waste stabilisation ponds; (4) Anaerobic reactors; (5) Activated sludge and aerobic biofilm reactors; (6) Sludge treatment and disposal.*

Volume 1 (*Wastewater characteristics, treatment and disposal*) presents an integrated view of water quality and wastewater treatment, analysing wastewater characteristics (flow and major constituents), the impact of the discharge into receiving water bodies and a general overview of wastewater treatment and sludge treatment and disposal. Volume 1 is more introductory, and may be used as teaching material for undergraduate courses in Civil Engineering, Environmental Engineering, Environmental Sciences and related courses.

Volume 2 (*Basic principles of wastewater treatment*) is also introductory, but at a higher level of detailing. The core of this book is the unit operations and processes associated with biological wastewater treatment. The major topics covered are: microbiology and ecology of wastewater treatment; reaction kinetics and reactor hydraulics; conversion of organic and inorganic matter; sedimentation; aeration. Volume 2 may be used as part of postgraduate courses in Civil Engineering, Environmental Engineering, Environmental Sciences and related courses, either as part of disciplines on wastewater treatment or unit operations and processes.

Volumes 3 to 5 are the central part of the series, being structured according to the major wastewater treatment processes (*waste stabilisation ponds, anaerobic reactors, activated sludge and aerobic biofilm reactors*). In each volume, all major process technologies and variants are fully covered, including main concepts, working principles, expected removal efficiencies, design criteria, design examples, construction aspects and operational guidelines. Similarly to Volume 2, volumes 3 to 5 can be used in postgraduate courses in Civil Engineering, Environmental Engineering, Environmental Sciences and related courses.

Volume 6 (*Sludge treatment and disposal*) covers in detail sludge characteristics, production, treatment (thickening, dewatering, stabilisation, pathogens removal) and disposal (land application for agricultural purposes, sanitary landfills, landfarming and other methods). Environmental and public health issues are fully described. Possible academic uses for this part are same as those from volumes 3 to 5.

Besides being used as textbooks at academic institutions, it is believed that the series may be an important reference for practising professionals, such as engineers, biologists, chemists and environmental scientists, acting in consulting companies, water authorities and environmental agencies.

The present series is based on a consolidated, integrated and updated version of a series of six books written by the authors in Brazil, covering the topics presented in the current book, with the same concern for didactic approach and balance between theory and practice. The large success of the Brazilian books, used at most graduate and post-graduate courses at Brazilian universities, besides consulting companies and water and environmental agencies, was the driving force for the preparation of this international version.

In this version, the books aim at presenting consolidated technology based on worldwide experience available at the international literature. However, it should be recognised that a significant input comes from the Brazilian experience, considering the background and working practice of all authors. Brazil is a large country with many geographical, climatic, economical, social and cultural contrasts, reflecting well the reality encountered in many countries in the world. Besides, it should be mentioned that Brazil is currently one of the leading countries in the world on the application of anaerobic technology to domestic sewage treatment, and in the post-treatment of anaerobic effluents. Regarding this point, the authors would like to show their recognition for the Brazilian Research Programme on Basic Sanitation (PROSAB), which, through several years of intensive, applied, cooperative research has led to the consolidation of anaerobic treatment and aerobic/anaerobic post-treatment, which are currently widely applied in full-scale plants in Brazil. Consolidated results achieved by PROSAB are included in various parts of the book, representing invaluable and updated information applicable to warm climate regions.

Volumes 1 to 5 were written by the two main authors. Volume 6 counted with the invaluable participation of Cleverson Vitorio Andreoli and Fernando Fernandes, who acted as editors, and of several specialists, who acted as chapter authors: Aderlene Inês de Lara, Deize Dias Lopes, Dione Mari Morita, Eduardo Sabino Pegorini, Hilton Felício dos Santos, Marcelo Antonio Teixeira Pinto, Maurício Luduvice, Ricardo Franci Gonçalves, Sandra Márcia Cesário Pereira da Silva, Vanete Thomaz Soccol.

Many colleagues, students and professionals contributed with useful suggestions, reviews and incentives for the Brazilian books that were the seed for this international version. It would be impossible to list all of them here, but our heartfelt appreciation is acknowledged.

The authors would like to express their recognition for the support provided by the Department of Sanitary and Environmental Engineering at the Federal University of Minas Gerais, Brazil, at which the two authors work. The department provided institutional and financial support for this international version, which is in line with the university's view of expanding and disseminating knowledge to society.

Finally, the authors would like to show their appreciation to IWA Publishing, for their incentive and patience in following the development of this series throughout the years of hard work.

Marcos von Sperling
Carlos Augusto de Lemos Chernicharo

December 2006

The author

Marcos von Sperling
PhD in Environmental Engineering (Imperial College, Univ. London, UK). Associate professor at the Department of Sanitary and Environmental Engineering, Federal University of Minas Gerais, Brazil. Consultant to governmental and private companies in the field of water pollution control and wastewater treatment. marcos@desa.ufmg.br

1

Overview of stabilisation ponds

The stabilisation pond systems constitute the simplest form of wastewater treatment. There are several variants of the stabilisation pond systems, with different levels of operational simplicity and land requirements. The following pond systems, whose main objective is the removal of carbonaceous matter, are covered in this part of the book:

- *Facultative ponds*
- *Anaerobic ponds followed by facultative ponds*
- *Facultative aerated lagoons*
- *Complete-mix aerated lagoons followed by sedimentation ponds*

Besides these ponds, *maturation ponds*, which may be included for the removal of pathogenic organisms, are also analysed.

However, in the present part of the book, only the ponds mentioned above are analysed in greater detail.

In general, stabilisation ponds are highly recommended for warm-climate areas and developing countries, due to the following aspects:

- sufficient land availability in a large number of locations
- favourable climate (high temperature and sunlight)
- simple operation
- little or no equipment required

Table 1.1. Brief description of the main stabilisation pond systems

System	Description
Facultative pond	The soluble and fine particulate BOD is aerobically stabilised by bacteria that grow dispersed in the liquid medium, while the BOD in suspension tends to settle, being converted anaerobically by bacteria at the bottom of the pond. The oxygen required by the aerobic bacteria is supplied by algae through photosynthesis. The land requirements are high.
Anaerobic pond – facultative pond	Around 50 to 70% of the BOD is converted in the anaerobic pond (deeper and with a smaller volume), while the remaining BOD is removed in the facultative pond. The system occupies an area smaller than that of a single facultative pond.
Facultative aerated lagoon	The BOD removal mechanisms are similar to those of a facultative pond. However, oxygen is supplied by mechanical aerators instead of through photosynthesis. The aeration is not sufficient to keep the solids in suspension, and a large part of the sewage solids and biomass settles, being decomposed anaerobically at the bottom.
Complete-mix aerated lagoon – sedimentation pond	The energy introduced per unit volume of the pond is high, which causes the solids (principally the biomass) to remain dispersed in the liquid medium, in complete mixing. The resulting higher biomass concentration in the liquid medium increases the BOD removal efficiency, which allows this pond to have a volume smaller than that of a facultative aerated lagoon. However, the effluent contains high levels of solids (bacteria) that need to be removed before being discharged into the receiving body. The sedimentation pond downstream provides conditions for the removal of these settleable solids. The sludge of the sedimentation pond must be removed every few years.
Maturation ponds	The main objective of maturation ponds is the removal of pathogenic organisms. In maturation ponds prevail environmental conditions which are adverse to these organisms, such as ultraviolet radiation, high pH, high DO, lower temperature (compared with the human intestinal tract), lack of nutrients and predation by other organisms. Maturation ponds are a post-treatment stage for BOD-removal processes, being usually designed as a series of ponds or a single-baffled pond. The coliform removal efficiency is very high.

Table 1.1 presents a brief description of the main pond systems analysed in this part of the book, while Table 1.2 compares some basic characteristics of the systems. The corresponding flowsheets are presented in Figures 1.1 and 1.2.

It should be noticed that the ponds can work as *post-treatment* for effluents from *anaerobic reactors* (such as *UASB* – Upflow anaerobic sludge blanket). When the removal of pathogenic organisms is the main objective, these post-treatment ponds are also called *polishing ponds* (see Figure 1.3), but they are basically maturation ponds, and their design parameters are very similar to those adopted for maturation ponds. If aerated lagoons are adopted as post-treatment, the detention time can be reduced, as a result of the lower input of organic matter load to the pond.

Table 1.2. Characteristics of the main pond systems

General item	Specific item	System of ponds				
		Facultative	Anaerobic–facultative	Facultative aerated	Compl.-mix aerated – sedim.	Anaerobic – facultative – maturation
Removal	BOD	75–85	75–85	75–85	75–85	80–85
Efficiency	COD	65–80	65–80	65-80	65–80	70–83
(%)	SS	70–80	70–80	70–80	80–87	73–83
	Ammonia	<50	<50	<30	<30	50–65
	Nitrogen	<60	<60	<30	<30	50–65
	Phosphorus	<35	<35	<35	<35	> 50
	Coliforms	90–99	90–99	90–99	90–99	99.9–99.9999
Requirements	Area (m^2/inhab.)	2.0–4.0	1.2–3.0	0.25–0.5	0.2–0.4	3.0–5.0
	Power (W/inhab.)	≈ 0	≈ 0	1.2–2.0	1.8–2.5	≈ 0
Costs	Construction	15–30	12–30	20–35	20–35	20–40
(US$/inhab)	O & M	0.8–1.5	0.8–1.5	2.0–3.5	2.0–3.5	1.0–2.0

Notes: O&M: operation and maintenance
Costs based on Brazilian experience

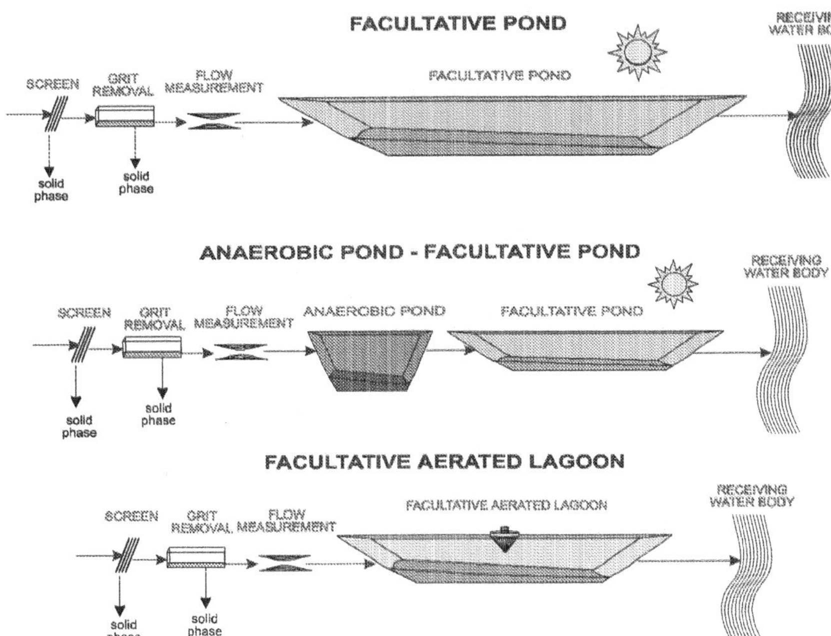

Figure 1.1.   Flowsheets of the main stabilisation pond systems applied for the removal of BOD

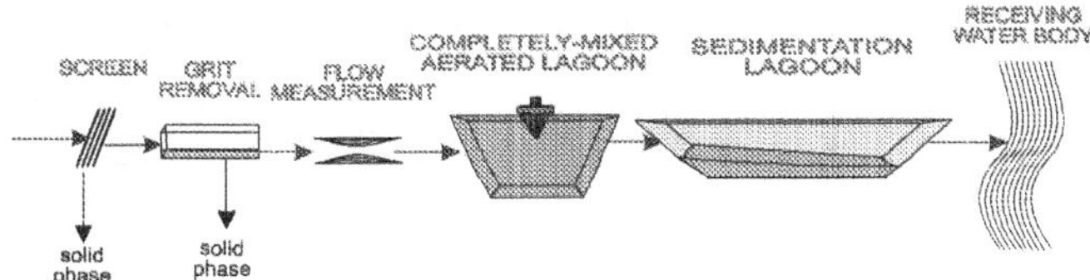

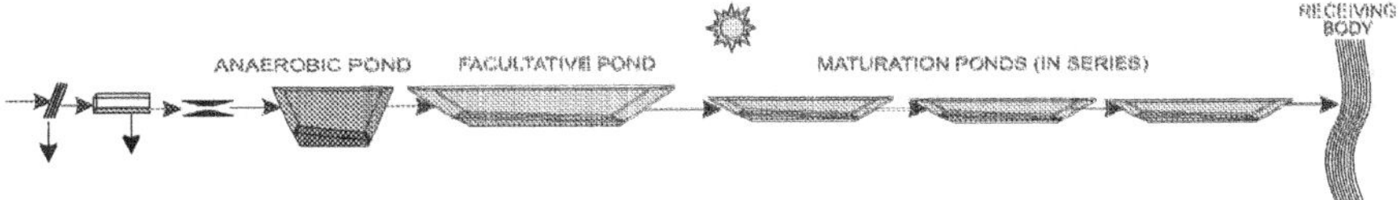

Figure 1.2.   Flowsheet of a system of stabilisation ponds followed by maturation ponds in series

Table 1.3. Typical removal efficiencies of pathogenic and indicator organisms in stabilisation pond systems

	Typical removal efficiency (% or log units removed) (*)				
Parameter	Facultative	Anaerobic – facultative	Facultative – maturation	Anaerobic – facultative – maturation	UASB reactor – polishing pond
Coliforms	1–2 log	1–2 log	3–6 log	3–6 log	3–6 log
Pathogenic bacteria	1–2 log	1–2 log	3–6 log	3–6 log	3–6 log
Viruses	≤ 1 log	≈ 1 log	2–4 log	2–4 log	2–4 log
Protozoan cysts	$\approx 100\%$	$\approx 100\%$	100%	100%	100%
Helminth eggs	$\approx 100\%$	$\approx 100\%$	100%	100%	100%

(*) 1 log = 90%; 2 log = 99%; 3 log = 99.9%; 4 log = 99.99%; 6 log = 99.9999%

Table 1.4. Sludge management in stabilisation ponds

Item	Anaerobic	Primary facultative	Secondary facultative	Maturation
Sludge accumulation rate (m^3/inhab.year)	0.02–0.10	0.03–0.09	0.03–0.05	–
Removal interval (years)	< 7	> 15	> 20	> 20
Total solids concentration in the sludge (% TS)	> 10% (c)	> 10% (c)	> 10% (c)	–
VS/TS ratio	< 50%	< 50%	< 50%	–
Coliform concent. in the sludge (FC/gTS)	10^2–10^4	10^2–10^4	10^2–10^4	10^2–10^4
Helminth eggs concent. in the sludge (eggs/gTS)	10^1–10^3	10^1–10^3	10^1–10^3	10^1–10^3
Additional treatment required	Dewat. (a)	Dewat. (a)	Dewat. (a)	–
Usual disposal routes	(b)	(b)	(b)	–

Obs: prior grit removal is essential
(a) Disinfection (usually lime treatment) in the case of agricultural use of the sludge
(b) Final disposal routes similar to those used for the other wastewater treatment processes (agricultural reuse, landfill, others)
(c) When removed by pumping, the concentration can decrease to values of 5 to 7%

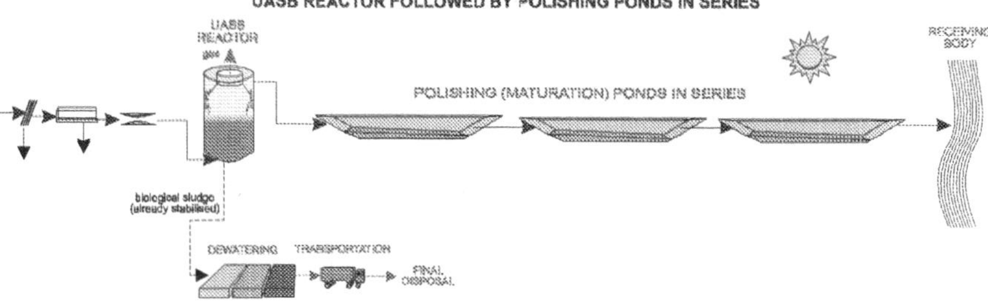

Figure 1.3.   Flowsheet of a system of UASB reactor followed by polishing (maturation) ponds in series

Table 1.5. Typical design parameters for stabilisation pond systems

Design parameter	Anaerobic	Facultative	Facultative aerated	Completely mixed aerated	Sedimentation	Maturation
Detention time t (d)	3–6	15–45	5–10	2–4	$\approx$2	(*)
Surface loading rate L_s (kgBOD$_5$/ha.d)	–	100–350	–	–	–	–
Volumetric loading rate L_V (kgBOD$_5$/m^3.d)	0.10–0.35	–	–	–	–	–
Depth H (m)	3.0–5.0	1.5–2.0	2.5–4.0	2.5–4.0	3.0–4.0	0.8–1.2
L/B ratio (length/breadth)	1 to 3	2 to 4	2 to 4	1 to 2	–	(**)
BOD removal coefficient K (complete mix) (20 °C) (d^{-1})	–	0.25–0.40	0.6–0.8	1.0–1.5	–	–
Temperature coefficient θ (complete mix)	–	1.05–1.085	1.035	1.035	–	–
BOD removal coefficient (dispersed flow) (20 °C)(d^{-1})	–	0.13–0.17	–	–	–	–
Temperature coefficient θ (dispersed flow)	–	1.035	–	–	–	–
Dispersion number d (L/B = 1)	–	0.4–1.3	–	–	–	0.4–1.1
Dispersion number d (L/B = 2 to 4)	–	0.1–0.7	–	–	–	0.1–0.5
Dispersion number d (L/B $\geq$ 5)	–	0.02–0.3	–	–	–	0.03–0.23
Effluent particulate BOD (mgBOD$_5$/mgSS)	–	0.3–0.4	0.3–0.4	0.3–0.6	–	–
Average O$_2$ requirements (kgO$_2$/kgBOD$_5$ remov)	–	–	0.8–1.2	1.1–1.4	–	–
Power level (W/m^3)	–	–	<2.0	$\geq$3.0	–	–
Coliform die-off coefficient K_b (complete mix) (20 °C) (d^{-1})	–	0.4–5.0	–	–	–	0.6–1.2 (***)
Temperature coefficient θ (complete mix)	–	1.07	–	–	–	1.07
Coliform die-off coefficient K_b (dispersed flow) (20 °C) (d^{-1})	–	0.2–0.3	–	–	–	0.4–0.7
Temperature coefficient θ (dispersed flow)	–	1.07	–	–	–	1.07

For details of the parameters: see text; L = length (m); B = breadth (m)
(*) The detention time in a maturation pond is a function of the pond shape and the required coliform removal efficiency
(**) L/B ratio including baffles in a single pond >10; L/B ratio in each pond of a series of more than 3 ponds: 1–5
(***) Coefficient K_b (complete mix) for maturation pond: values given are for ponds in series (baffled ponds are not well represented by the complete-mix model)

Regarding the removal of *pathogenic* organisms, a series of ponds *including maturation ponds* is capable of reaching very high removal efficiencies. Typical efficiencies of widely used pond systems for pathogen removal are presented in Table 1.3.

Sludge management in unaerated ponds is summarised in Table 1.4. Details are presented in the respective chapters, including the aerated lagoons. Sludge management is analysed specifically in Chapter 11.

A summary of the main design criteria adopted for the pond systems covered in this part of the book is presented in Table 1.5.

2

Facultative ponds

2.1 INTRODUCTION

Facultative ponds are the simplest variant of the stabilisation ponds systems. Basically, the process consists of the retention of wastewater for a period long enough, so that the natural organic matter stabilisation processes take place. Therefore, the main advantages and disadvantages of facultative ponds are associated to the predominance of natural phenomena.

The advantages are associated with the high operational simplicity and reliability. Natural processes are likely to be reliable: there is no equipment that can be out of order or the need for special operational schemes. However, nature is slow and needs long detention times so that the reactions are completed, which implies large land requirements. The biological activity is largely affected by temperature, mainly under the natural conditions of the ponds. As a result, the stabilisation ponds are more appropriate where the land is cheap, the climate is favourable, and a treatment method that does not require equipment or a special training for the operators is desired (Arceivala, 1981).

The costs of stabilisation ponds are very competitive, as long as the land costs or the need of earth works is not excessive. The construction is simple and involves mainly earth works, and the operational costs are much smaller than in other treatment methods. The efficiency of the system is usually satisfactory, and levels comparable to many secondary treatment systems can be obtained.

Figure 2.1 presents the typical flowsheet of a facultative pond.

A terminology frequently adopted for ponds is related to their position in the series of treatment units:

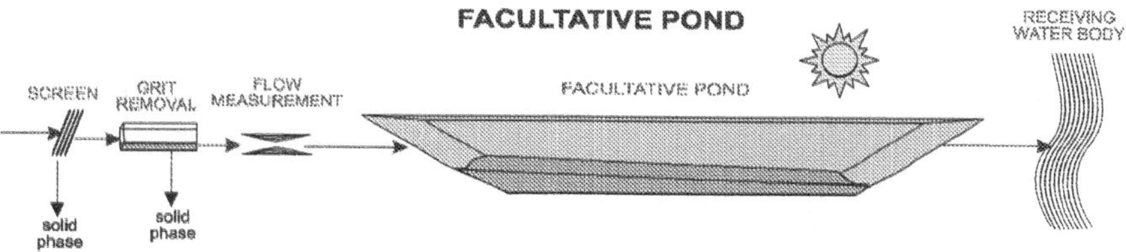

Figure 2.1.   Flowsheet of a facultative pond

- *Primary pond*: first pond of the series - facultative pond that receives raw sewage
- *Secondary pond*: second pond of the series - receives effluent from another unit upstream (usually an anaerobic pond)
- *Tertiary, quaternary ponds*, etc.: occupy the third, fourth, etc. position in the series – they are usually maturation ponds

2.2 DESCRIPTION OF THE PROCESS

The influent wastewater enters at one end of the pond and leaves at the opposite end. During this time, which takes several days, a series of mechanisms contribute to the purification of the wastewater. These mechanisms occur in three zones of the ponds, denominated: *anaerobic zone, aerobic zone* and *facultative zone.*

The suspended organic matter (*particulate BOD*) tends to settle, constituting the bottom sludge (**anaerobic zone**). This sludge undergoes a decomposition process by anaerobic microorganisms, being slowly converted into carbon dioxide, methane and others. After a certain period, practically only the inert fraction (non-biodegradable) remains in the bottom layer. The hydrogen sulphide generated does not cause malodour problems, since it is oxidised by chemical and biochemical processes in the upper aerobic layer.

The dissolved organic matter (*soluble BOD*), together with the small suspended organic matter (*finely particulate BOD*) does not settle and remains dispersed in the liquid mass. In the upper layer, an **aerobic zone** is present. In this zone, the organic matter is oxidised by aerobic respiration. Oxygen is required, which is supplied to the medium by the photosynthesis undertaken by algae, and there is a balance between the consumption and production of oxygen and carbon dioxide (see Figure 2.2):

<table>
<tr><td>

Bacteria → *respiration*:
- Consumption of oxygen
- Production of carbon dioxide

Algae → *photosynthesis*:
- Production of oxygen
- Consumption of carbon dioxide

</td></tr>
</table>

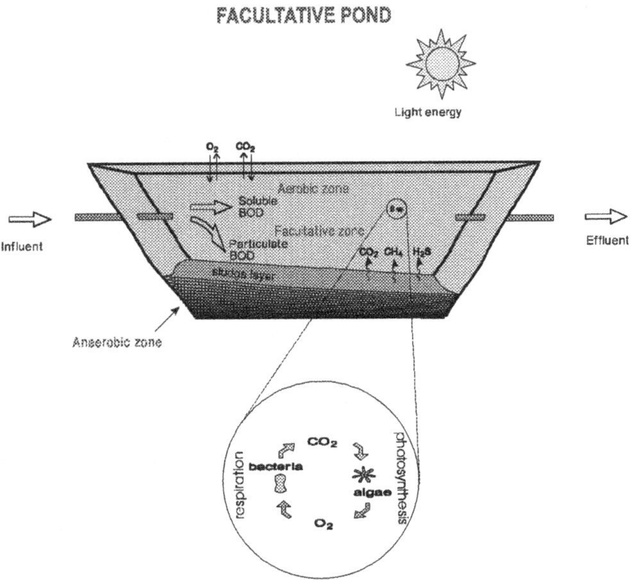

Figure 2.2.   Simplified working principle of a facultative pond

It should be highlighted that the reactions of *photosynthesis* (production of organic matter) and *respiration* (oxidation of the organic matter) are similar, but with opposite directions:

- Photosynthesis:
  $$CO_2 + H_2O + Energy \rightarrow organic\ matter + O_2$$
- Respiration:
  $$Organic\ matter + O_2 \rightarrow CO_2 + H_2O + Energy$$

For the occurrence of photosynthesis, a source of *light energy* is necessary, in this case, represented by the sun. For this reason, localities with high solar radiation and a low level of cloudiness are highly favourable for facultative ponds.

Since photosynthesis depends on solar energy, it reaches higher levels close to the pond surface. Deeper down in the pond, light penetration is smaller, which causes the predominance of the oxygen consumption (respiration) over its production (photosynthesis), with the occasional absence of dissolved oxygen from a certain depth. Besides, photosynthesis only occurs during the day (sunshine hours), and during the night, the absence of oxygen can prevail. Because of these facts, it is essential that there are several groups of bacteria, responsible for the stabilisation of the organic matter, which can survive and proliferate in the *presence* as well as in the *absence* of oxygen. In the absence of free oxygen, other electron acceptors are used, such as nitrates (anoxic conditions). This zone, where the presence or the absence of oxygen can occur, is called a **facultative zone**. This condition also gives the name to the ponds (facultative ponds).

As commented, the process of facultative ponds is essentially natural and does not need any equipment. For this reason, the stabilisation of the organic matter takes place at slow rates, implying the need of a high detention time in the pond (usually greater than 20 days). Photosynthesis, to be effective, requires a high exposure area for the best use of the solar energy by the algae, justifying the need of large units. Consequently, the total area required by facultative ponds is the largest amongst all the wastewater treatment processes (excluding land disposal systems). On the other hand, the fact that they are a natural process is associated with a larger operational simplicity, which is a factor of fundamental importance in developing countries.

The effluent from a facultative pond has the following main characteristics (CETESB, 1989):

- green colour due to the algae
- high dissolved oxygen concentration
- high suspended solids concentration, although these practically do not settle (the algae practically do not settle in the Imhoff-cone test)

2.3 INFLUENCE OF ALGAE

Algae play a fundamental role in facultative ponds. Their concentration is much higher than that of bacteria, giving the greenish appearance of the liquid at the pond surface. In terms of dry suspended solids, their concentration is usually lower than 200 mg/L, although in terms of numbers they can reach counts in the range of 10^4 to 10^6 organisms per ml (Arceivala, 1981). The presence of algae is usually measured in the form of chlorophyll a, a pigment presented by all plants, and the main parameter for the quantification of the algal biomass (König, 2000). The chlorophyll a concentrations in facultative ponds depend on the applied load and temperature, but are usually located in the range from 500 to 2000 µg/L (Mara et al, 1992).

The main types of algae found in stabilisation ponds are (Mara et al, 1992; Silva Jr. and Sasson, 1993; Jordão and Pessoa, 1995):

- **Green algae** (Chlorophyta) and **pigmented flagellated** (Euglenophyta). These algae give the pond the predominant greenish colour. The main genera are *Chlamydomonas*, *Chlorella* and *Euglena*. *Chlamydomonas* and *Euglena* are usually the first to appear in the pond, tending to be dominant in cold periods, and possessing flagella, which gives them motility (optimisation of their position with relation to the incidence of light and to temperature).
- **Cyanobacteria** (previously called Cyanophyta or blue-green algae). In reality these organisms present characteristics of bacteria and algae, and are classified as bacteria. The cyanobacteria do not have locomotion organelles,

such as cilia, flagella or pseudopodes, but are capable of moving by sliding. The nutrient requirements are very small: the cyanobacteria can proliferate in any environment that has at least CO_2, N_2, water, some minerals and light. These organisms are typical of conditions with low pH values and low nutrient availability in the wastewater. This environment (not typical in stabilisation ponds) is unfavourable for the green algae, which may also serve as food for other organisms, such as protozoa, leading to the proliferation of the cyanobacteria. *Oscillatoria*, *Phormidium*, *Anacystis* and *Anabaena* are among the main genera that can be mentioned.

Other types that can be found are algae of the phyla Bacyllariophyta and Chrysophyta (König, 2000; Mara et al, 1992). The predominant species vary from place to place, and even with the position in the series of ponds (facultative ponds and maturation ponds).

The algae photosynthesise during the hours of the day that are subject to light radiation. In this period, they produce the organic matter necessary for their survival, converting the light energy into condensed chemical energy in the form of food. During the 24 hours of the day, they respire, oxidising the organic matter produced, and release the energy for growth, reproduction, locomotion and others. The balance between oxygen production (photosynthesis) and consumption (respiration) widely favours the former. In fact, the algae may produce about 15 times more oxygen than they consume (Abdel-Razik, 1991), leading to a positive balance of DO in the system.

Owing to the requirement of light energy, most of the algae are located close to the pond surface, a location of high oxygen production. When deepening down into the pond, the light energy decreases, therefore reducing the algal concentration. In the surface layer, under 50 cm, is the range of higher light intensity, with the rest of the pond being practically dark.

There is a position in the pond depth in which the oxygen production by the algae equals the oxygen consumption by the algae and the decomposing microorganisms. This point is called **oxypause** (see Figure 2.3).

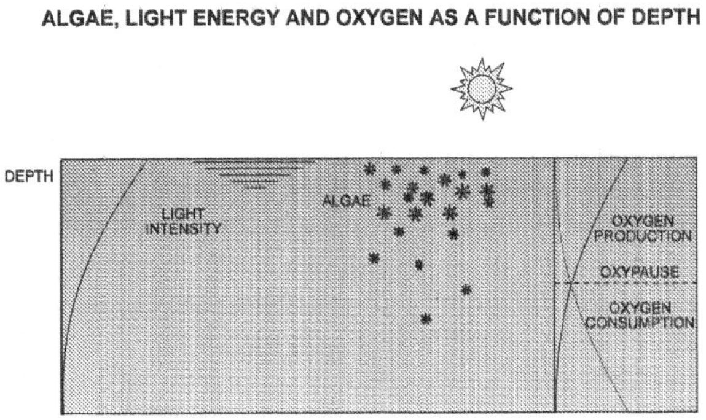

Figure 2.3.   Algae, light energy and oxygen in a facultative pond (cross-section)

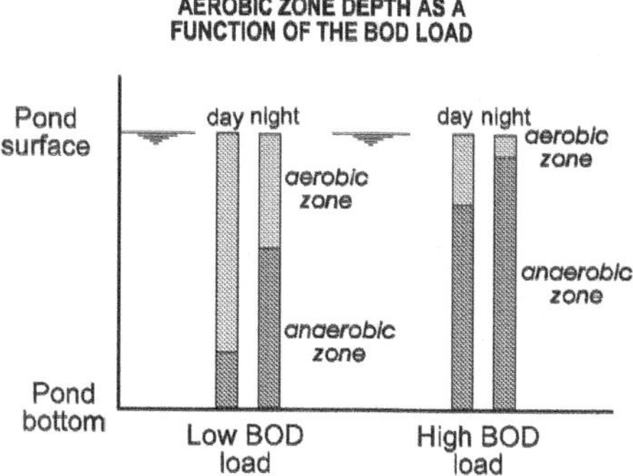

Figure 2.4.   Influence of the organic load applied to the pond and the hour of the day on the thickness of the aerobic and anaerobic layers (adapted from Arceivala, 1981)

Above the oxypause, aerobic conditions prevail, while below it, anoxic or anaerobic conditions predominate. The level of the oxypause varies during the 24 hours of the day, as a function of the variability of the photosynthesis during this period. At night, the oxypause level rises in the pond, while during the day it lowers down.

The thickness of the aerobic zone, besides varying along the day, also varies with the loading conditions of the pond. Ponds with a greater BOD load tend to have a larger anaerobic layer, which can practically take up the whole pond depth during the night. Figure 2.4 schematically illustrates the influence of the loading conditions on the thickness of the aerobic layer.

The pH in the pond also varies with the depth and along the day. The pH depends on the photosynthesis and respiration, according to:

- *Photosynthesis:*
    - Consumption of CO_2
    - Bicarbonate ion (HCO_3^-) of the wastewater is converted to OH^-
    - pH rises
- *Respiration:*
    - Production of CO_2
    - Bicarbonate ion (HCO_3^-) of the wastewater is converted to H^+
    - pH decreases

During the day, in the hours of maximum photosynthetic activity, the pH can reach values around 9 or even more. In these conditions of high pH, the following phenomena can occur:

- Conversion of the ammonium ion (NH_4^+) to free ammonia (NH_3), which is toxic, but tends to be released to the atmosphere (nutrient removal)
- Precipitation of the phosphates (nutrient removal)
- Conversion of sulphide (H_2S), which may cause bad odours, to the odourless bisulphide ion (HS^-). At pH levels greater than 9 there is practically no H_2S.

Table 2.1. Influence of the main external environmental factors

Factor	Influence
Solar radiation	• Photosynthesis velocity
Temperature	• Photosynthesis velocity • Bacterial decomposition rate • Gas solubility and transfer • Mixing conditions
Wind	• Mixing conditions • Atmospheric reaeration (*)

(*) Mechanism of lesser importance in the DO balance in facultative ponds

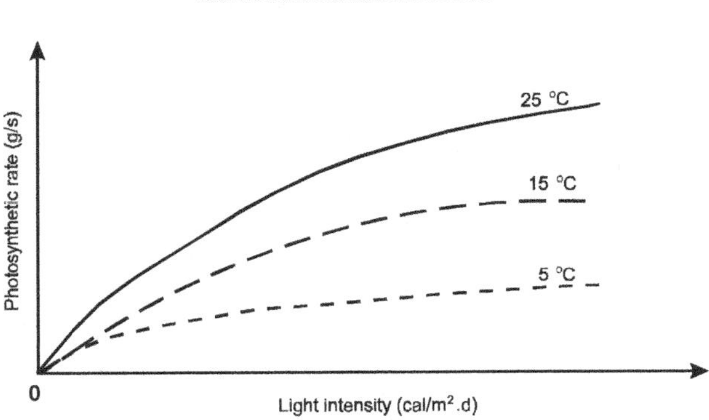

Figure 2.5.   Influence of temperature and light radiation in the photosynthetic velocity (adapted from Jordão and Pessôa, 1995)

2.4 INFLUENCE OF ENVIRONMENTAL CONDITIONS

The main environmental conditions in a stabilisation pond are *solar radiation*, *temperature* and *wind* – see Table 2.1 (Jordão and Pessôa, 1995).

The influence of the temperature and solar radiation in the photosynthetic rate is shown schematically in Figure 2.5.

a) Mixing and thermal stratification

Mixing in a stabilisation pond occurs mainly through the following mechanisms: wind and temperature difference. Mixing is important for the performance of the pond due to the following beneficial aspects (Silva and Mara, 1979):

• Minimisation of the occurrence of hydraulic short circuits
• Minimisation of the occurrence of stagnant zones (dead zones)
• Homogenisation of the vertical distribution of BOD, algae and oxygen
• Transport to the photic surface zone of non-motile algae that would tend to settle

STRATIFICATION AND MIXING DYNAMICS IN PONDS

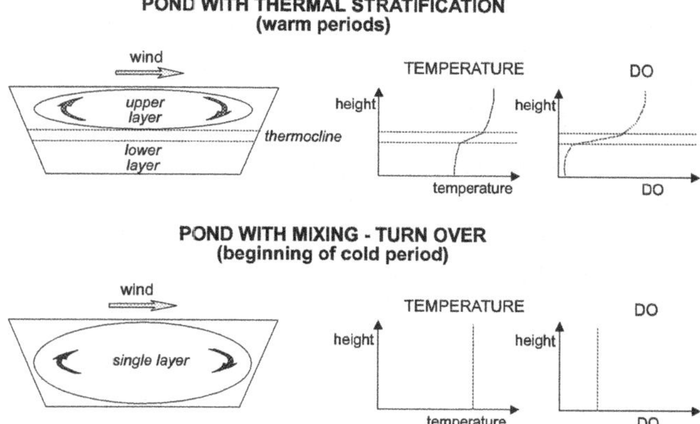

Figure 2.6. Stratification and mixing in a pond

- Transport to the deeper layers of the oxygen produced by photosynthesis in the photic zone

To maximise the influence of the *wind*, the pond should not be surrounded by natural or artificial obstacles that could obstruct the wind access. Additionally, the pond should not have a very irregular shape, which could hinder the homogenisation of the peripheral areas with the main pond body.

The pond is also subject to *thermal stratification*, in which the upper layer (warm) is not mixed with the lower (cold) layer. When deepening down in the pond, there is a point with a great decrease in the temperature, accompanied by high density and viscosity increases. This point is called the *thermocline*. Thus, two distinct layers are formed: the superficial one (lower density) and the bottom one (greater density), which are not mixed (see Figure 2.6).

The behaviour of the algae is influenced by the stratification according to:

- The non-motile algae settle and reach the dark zone of the pond, where they stop producing oxygen, leading, on the other hand, only to its consumption.
- The motile algae tend to escape from the upper surface layer (30 to 50 cm) of high temperature (occasionally 35 °C or more), and form a dense layer of algae, which hinders the penetration of the solar energy.

Because of these aspects, in stratified ponds there may be a low presence of algae in the photic zone, which reduces the oxygen production of the system and consequently its capacity to stabilise the organic matter. In locations with little or no wind at the pond surface, the pond may remain stratified.

The stratification can be interrupted by means of a natural mixing mechanism, denominated *turnover* or *thermal inversion* (see Figure 2.6). In stratified tropical lakes, the thermal inversion can take place in the cold period (winter). Besides this, in shallow lakes, such as stabilisation ponds, the mixing can happen once a day, according to the following sequence (Silva and Mara, 1979):

- *Beginning of the morning, with wind.* Complete mixing. The temperature is uniform throughout the depth.
- *Middle of the morning, with sun, without wind.* Increase of the temperature in the surface layer (above the thermocline). Little variation of the temperature at the bottom (below the thermocline), which is influenced by the ground temperature. Stratification.
- *Beginning of the night, without wind.* The layer above the thermocline loses heat more quickly than the bottom layer. If the temperatures of the layers become similar, mixing occurs.
- *Night, with wind.* The wind aids in the mixing of the layers. The upper layer sinks and the bottom layer rises.

Figure 2.7 shows experimental results (mean values) of temperature in a shallow pilot pond (1.0 m deep, with baffles, length/breadth ratio = 32), located in South-

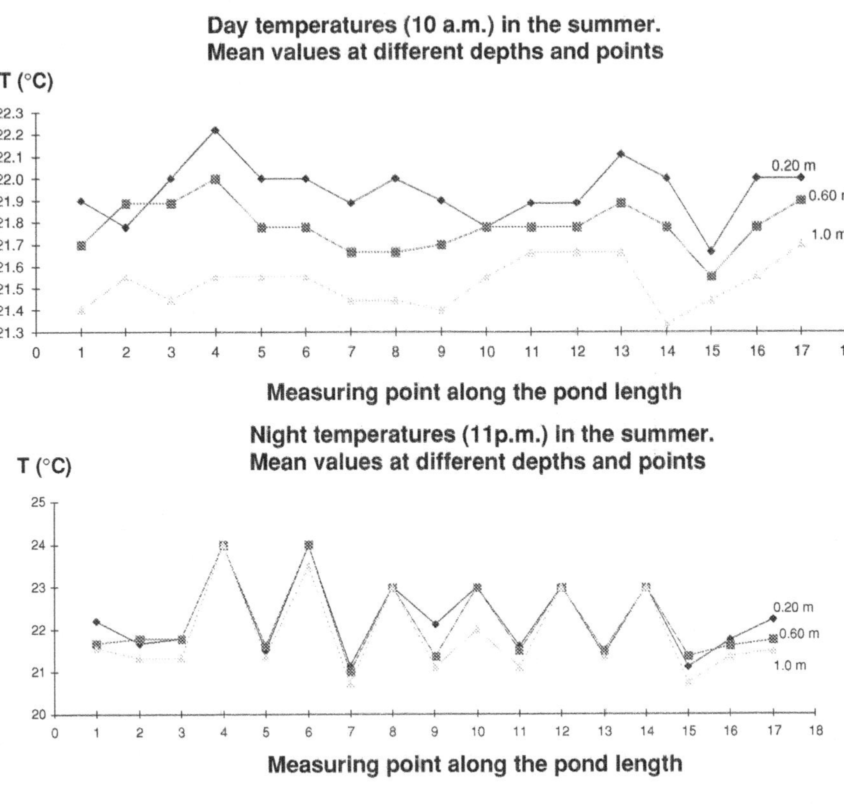

Figure 2.7.   Longitudinal profile of the temperature in a pilot baffled pond, at daily and nightly hours. Measurements at the depths of 0.20 m, 0.60 m, and 1.00 m below the water level. Pond depth: 1.00 m.

east of Brazil. The measurements were made at depths of 0.2 m, 0.6 m, and 1.0 m below the water level, and along the longitudinal course of the liquid in the pond. The figure shows summer data taken at 10 a.m., clearly indicating stratification (higher temperatures at the shallower depths, closer to the water level). However, at 11 p.m., also in the summer, the pond becomes totally mixed. The winter data is not presented here, but they indicate total mixing in the morning as well as in the night.

Kellner and Pires (1998, 1999) present a mathematical model for the estimation of thermal stratification in stabilisation ponds. They point out that the stratification leads to a loss of the net volume of the pond, and that the volume of the upper layer may be insufficient for the completion of the desired biochemical reactions.

b)   Relationship between the air and the liquid temperature

The *average temperature of the liquid in the coldest month* is usually considered in many designs. Yanez (1993) and Brito et al (2000) present correlation studies between the air and the liquid temperature, in two ponds in Brazil, two in Peru and one in Jordan. The regressions are presented in Figure 2.8. The figure also presents a straight line, calculated by the author, based on the average values of the five equations. The resulting equation is:

$$\boxed{T_{liquid} = 12.7 + 0.54 \times T_{air}}$$

(2.1)

Table 2.2 presents the resulting values of the water temperature calculated using Equation 2.1 for different values of the air temperature. The values obtained in the range of 20 to 30 °C are in agreement with the comment from Mara et al (1997) that the temperature of the pond is about 2 to 3 °C warmer than the temperature of the air in the cold period, the inverse occurring in the hot

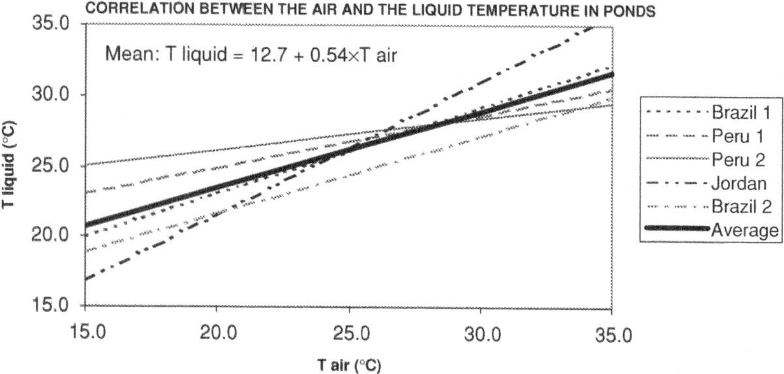

Figure 2.8.   Lines of best fit for the regressions between the water and the air temperatures in five ponds. Data from Yanez (1993) and Brito et al (2000). Average line calculated by the author.

Table 2.2. Water temperature in the pond, as a function of
the air temperature

Air temperature (°C)	Average liquid temperature (°C)
15	20.8
20	23.5
25	26.2
30	28.9
35	31.6

Estimation of the liquid temperature using Equation 2.1

period. An additional interpretation of the data from Yanez (1993) leads to the
conclusion that the temperature in the surface of the pond is 1 to 5 °C higher
than the average temperature, with the largest differences occurring in the warm
period.

2.5 DESIGN CRITERIA

The main parameters for the design of facultative ponds are:

- Surface organic loading rate
- Depth
- Detention time
- Geometry (length / breadth (L/B) ratio)

Surface organic loading rate. The surface organic loading rate (organic load
per unit area) is the main design criterion for facultative ponds. It is based on the
need to have a certain exposure area to the sun light in the pond, so that the process of
photosynthesis may take place. The objective of guaranteeing photosynthesis and
algal growth is to have enough oxygen production to counterbalance the oxygen
demand. Thus, the surface loading rate criterion is associated with the need of
oxygen for the stabilisation of the organic matter. Therefore, the surface loading
rate is related to the activity of *algae* and the balance between oxygen production
and consumption.

Depth. The depth has an influence on the physical, biological, and
hydrodynamic aspects of the pond. After obtaining the value of the surface area
(through the adoption of a value for the surface loading rate) and adopting a value
for the depth, the volume of the pond is obtained.

Detention time. The detention time is not a direct design parameter, but a verification parameter (resulting from the determination of the pond volume). The detention time criterion is associated with the time necessary for the microorganisms
to stabilise the organic matter in the reactor (pond). Therefore, the detention time
is related to the activity of the *bacteria*.

Pond geometry. The length to breadth (L/B) ratio is another important criterion, since it affects the hydraulic regime in the pond, which can be designed to
approximate plug-flow or complete-mix conditions.

The design parameters are basically empirical. For the surface loading rate, there are some mathematical models that allow the design of facultative ponds based on conceptual methods, such as algae production as a function of the solar radiation, oxygen production per unit algal mass and others. However, such methods are outside the scope of the present book, where the approach is essentially simplified. Besides this, the empirical methods have been traditionally used, based on experience acquired in several areas of the world.

a)   Surface organic loading rate

The area required for the pond is calculated as a function of the surface loading rate L_s. The rate is expressed in terms of the BOD load (L, expressed in $kgBOD_5/d$) that can be treated per unit surface area of the pond (A, expressed in ha).

$$A = L/L_s \qquad (2.2)$$

where:
   A = area required for the pond (ha)
   L = influent total (soluble + particulate) BOD ($kgBOD_5/d$)
   L_s = surface loading rate ($kgBOD_5/ha.d$)

The rate to be adopted varies with the local temperature, latitude, solar exposure, altitude and others. Locations with extremely favourable climate and sunlight allow the adoption of very high rates, occasionally greater than 300 $kgBOD_5/ha.d$, which implies smaller surface areas. On the other hand, temperate climate locations require loading rates lower than 100 $kgBOD_5/ha.d$. In *tropical and subtropical-climate regions*, the following rates have been adopted:

• Regions with warm winter and high sunshine: $L_s = 240$ to 350 $kgBOD_5/ha.d$
• Regions with moderate winter and sunshine: $L_s = 120$ to 240 $kgBOD_5/ha.d$
• Regions with cold winter and low sunshine: $L_s = 100$ to 180 $kgBOD_5/ha.d$

There are several empirical equations available on the international literature, correlating the surface loading rate L_s with the temperature T. One of the equations, proposed by Mara (1997), is presented below. According to him, the equation has global applicability. The equation uses the **mean temperature of the air in the coldest month**. The reason for using the mean temperature of the air is that, in the cold period, a safe value is obtained, since the temperature of the water will be slightly higher. The selection of the cold period is because it is the most critical in the operation of the pond, in terms of the velocities of the biochemical reactions. In the design of the facultative ponds in this book, the mean temperature of the liquid in the coldest month is adopted (in order to calculate the BOD removal rates). However, to estimate the surface loading rate, the safe assumption proposed by Mara is adopted (that is, to consider the air temperature the same as the liquid temperature). Section 2.4.b discusses the relationship between the water and the air temperature.

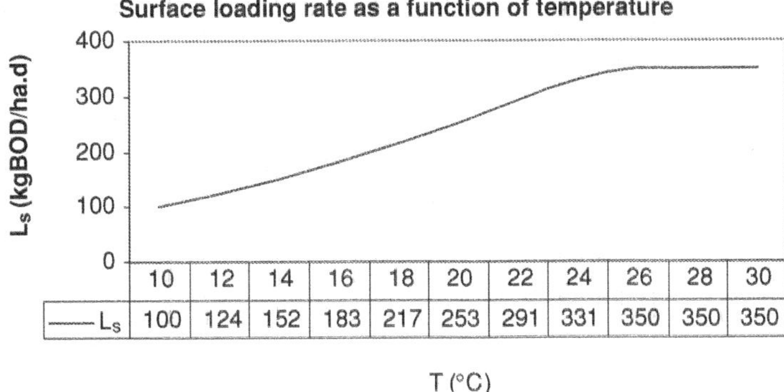

	10	12	14	16	18	20	22	24	26	28	30
—— L_s	100	124	152	183	217	253	291	331	350	350	350

Figure 2.9. Values of the surface loading rate as a function of the mean air temperature in the coldest month (according to Equation 2.3, Mara, 1997)

$$L_s = 350 \times (1.107 - 0.002 \times T)^{(T-25)} \qquad (2.3)$$

where:

T = mean air temperature in the coldest month (°C)

The application of Equation 2.3 produces the values of L_s presented in Figure 2.9. Even though Equation 2.3 leads to very high values of L_s with high temperatures (above 25 °C), it is recommended that the surface loading rate be limited to a maximum value of 350 kgBOD/ha.d for design purposes.

Naturally, the use of an empirical formula is only for an initial estimate of the surface loading rate. As commented, if there are local experiences, as well as other climatic evidences that suggest the adoption of other values, these specificities should always be taken into consideration when selecting the value of L_s.

There is no absolute maximum value for the surface area, beyond which facultative pond systems become unfeasible. The desirability of adopting more compact systems if large ponds are required depends essentially on the local conditions, topography, geology and land cost. Similarly, the division of a single pond into ponds in parallel depends on topography and the desirability to have more flexibility and improved hydraulics.

b)   Depth

As seen, the aerobic zone of the facultative pond depends on the penetration of sun light to give support to the photosynthetic activity. The intensity of light in the water body tends to reduce exponentially with depth. This phenomenon occurs even in distilled water, although at a much lower magnitude. The larger the colour and turbidity of the water and its algae concentration, the faster the light extinguishes. Below a certain depth in the pond, the environment is inappropriate for the growth of algae.

Table 2.3. Aspects related to the pond depth

Depth	Aspect
Shallow	• Shallow ponds, with depths lower than 1.0 m, can be completely aerobic. • The required area is very high, in order to comply with the detention time requirement. • The penetration of light through the depth is practically complete (the light energy tends to extinguish with depth, even in clean waters). • The production of algae is maximised and the pH is usually high (due to photosynthesis), causing the precipitation of phosphates and the stripping of ammonia (removal of nutrients). • Due to the low depth, there can be the development of emergent vegetation, which is a potential shelter for mosquito larvae (ponds with a depth around 0.60 m or less). • Shallow ponds are more affected by ambient temperature variations along the day, and can reach anaerobic conditions in warm periods (increase of the decomposition rate of the organic matter and a larger influence of the resolubilisation of by-products from the anaerobic decomposition of the sludge at the bottom).
Deep	• Ponds with higher depths provide a larger detention time for the stabilisation of the organic matter. • The performance of the pond is more stable and less affected by environmental conditions, producing an effluent with a more uniform quality throughout the year. • There is a larger storage volume for the sludge. • The bottom layer stays in anaerobic conditions, in which the BOD removal rate and the pathogenic death rate are slower. • The anaerobic decomposition obviously does not consume the dissolved oxygen in the medium. Thus, in the calculation of the DO balance, the fraction of the organic matter subject to the anaerobic decomposition can be taken into consideration. Usually, for a question of safety, the total influent BOD is considered to exert the oxygen demand, and for that the photosynthetic production in the upper layer should be sufficient. • The by-products of the anaerobic decomposition are released to the upper layers, still exerting some oxygen demand. The risks of bad smells are reduced, because in the aerobic layer the sulphide generated in the anaerobic decomposition is oxidised chemically and biochemically. • The deeper ponds allow future expansion for the inclusion of aerators, becoming aerated lagoons.

Based on the area and volume criteria, the depth H of the pond is a compromise between the required volume V and the required area A, considering that H = V/A. However, other aspects influence the selection of the depth of the pond (Arceivala, 1981), as listed in Table 2.3.

In conclusion, the available knowledge is still limited to optimise the depth of the pond, in order to maximise the number of benefits. The depth range to be adopted in the design of facultative ponds lies between 1.5 to 3.0 m, although the following range is more usual:

$$H = 1.5 \text{ m to } 2.0 \text{ m}$$

c) Detention time

The detention time of the pond is associated with the volume and the design flow:

$$\boxed{t = V/Q}$$
(2.4)

where:
t = detention time (d)
V = pond volume (m^3)
Q = average influent flow (m^3/d)

The average flow is the average of the influent flow and the effluent flow. The effluent flow corresponds to the influent flow minus the sinks plus the sources:

$$Q_{average} = (Q_{infl} - Q_{effl})/2$$
(2.5)

$$Q_{effl} = Q_{infl} + Q_{precipitation} - Q_{evaporation} - Q_{infiltration}$$
(2.6)

The additional components in Equation 2.6 can usually be ignored. For example, in a location where the average annual precipitation is 1,000 mm/year, the evaporation is 2,000 mm/year, the influent flow is 3,000 m^3/d (1,095,000 m^3/year) and the surface area of the pond is 48,000 m^2 (flow and area of Example 2.3), one has (ignoring the infiltration):

$$Q_{effl} = (1,095,000 \, m^3 \, year) + (1.0 \, m/year \times 48,000 \, m^2)$$
$$- (2.0 \, m/year \times 48,000 \, m^2)$$
$$= 1,095,000 + 48,000 - 96,000 = 1,047,000 \, m^3/year$$

In this case, the annual loss is only 4.4% of the influent flow. However, depending on the circumstances, in certain dry months there may not be rainfall, at the same time that there is a substantial evaporation rate. In these cases, the water balance may be affected, and the loss (or occasional gain, in an opposite situation) can be more significant. Infiltration may also play an important role, especially in ponds with unsealed bottoms (see Chapter 9).

The detention time required for the oxidation of the organic matter varies with the local conditions, especially the temperature. In **primary** facultative ponds treating *domestic sewage*, the resulting detention times usually vary between:

$$\boxed{t = 15 \text{ to } 45 \text{ days}}$$

The lower detention times occur in areas where the liquid temperature is higher, and a reduction in the volume required for the pond is achieved. The required detention time is a function of the kinetics of the BOD removal and the hydraulic regime of the pond (see Section 2.6.1). In locations with concentrated sewage (low per capita sewage flow and a high BOD concentration), the detention time tends to be high.

With highly concentrated industrial wastewaters, the resulting detention time is usually much higher, because the pond area (and, indirectly, volume) is calculated based on organic load, and not on flow (which is comparatively low, for a given BOD load). The decisive factor, in the case of industrial effluents, continues to be the organic loading rate.

The surface loading rate and detention time criteria are complementary, that is, the area and the volume obtained should be coherent. The detention time can be used in one of the following two ways:

- *Adopt t as an explicit design parameter.* After t has been adopted, V is calculated (V = t.Q). Since the area A has been already determined based on the surface loading rate, the depth H can be calculated (H = V/A) and verified whether it is inside the range presented in Item b.
- *Adopt a value for the depth H, according to the criteria of Item b.* Having H and A, the volume V is calculated (V = A.H) and, in consequence, the detention time t (t = V/Q).

With the value of t, the effluent BOD concentration is estimated (see Section 2.6). If the effluent concentration does not satisfy the requirements, the volume, or the detention time, should be increased.

The second approach is more practical, because it adopts objective values for the surface area and depth. Example 2.3 shows the joint interpretation of these two criteria.

d) Geometry of the pond (length / breadth ratio)

As discussed in Section 2.6.1, the hydraulic regime of plug-flow is the most efficient in terms of the removal of constituents that follow first-order kinetics, such as the organic matter and coliforms. However, the complete-mix regime is more suitable when the wastewater is subject to highly variable loads and the presence of toxic compounds, due to the fact that complete-mix reactors provide an immediate dilution of the influent in the liquid mass (see Chapter 8 for further details).

Plug-flow reactors are also subject to a high oxygen demand close to the pond inlet, as a result of the arrival of raw wastewater, without dilution, in the body of the reactor. Anaerobic conditions can occur as a consequence of the localised organic overload (high organic loading rate in the inlet portion of the pond). For this reason, the following statements can be made:

- ***Primary facultative ponds:*** *not usually designed to approach plug-flow reactors* (high length/breadth ratio) with the introduction of baffles, due to the possibility of organic overload close to the pond inlet.
- ***Secondary facultative ponds:*** also not usually designed to approach plug-flow conditions, but there is *more flexibility in the selection of the L/B ratio.*

> • ***Maturation ponds:*** most of the organic matter has been already previously removed, and there is less concern with an overload in the initial compartment. This allows the adoption of elongated ponds or baffles, leading to *high L/B ratios*.

The design of ponds can make use of the available site and its topography to obtain the most adequate length/breadth (L/B) ratio. Systems with high L/B tend to plug flow, while ponds with L/B close to 1.0 (square ponds) approach complete-mix conditions. More frequently, the L/B ratio for facultative ponds is situated within the following range (EPA, 1983; Abdel-Razik, 1991):

> Length / breadth (L/B) ratio = 2 to 4

2.6 ESTIMATION OF THE EFFLUENT BOD CONCENTRATION

2.6.1 Influence of the hydraulic regime

BOD removal follows a *first-order reaction* (in which the reaction rate is directly proportional to the substrate concentration). Under these conditions, the hydraulic regime of the reactor (pond) influences the efficiency of the system.

Although the kinetics of BOD removal are the same in the different hydraulic regimes, the effluent BOD concentration varies. According to the first-order kinetics, the BOD removal rate is higher the greater is the BOD concentration in the medium. This aspect has a great implication in the performance of the reactor, as seen below:

- **Plug-flow reactors.** In reactors in which there is a high BOD concentration (for example, close to the inlet), the removal rate is higher at this point. This is the case, for instance, of predominantly longitudinal reactors, such as the plug-flow reactors (the concentration close to the reactor inlet is different from the effluent concentration).
- **Complete-mix reactors.** Reactors that allow an immediate dispersion of the pollutant as a result of the homogenisation of the entire tank cause the influent concentration to rapidly equal the low effluent concentration. The low concentrations prevailing in the reactor lead to a lower BOD removal efficiency. This is the case of predominantly square complete-mix reactors (the concentration in the reactor, close to the inlet, is equal to the concentration at the outlet).

These two types of *idealised reactors* characterise an envelope, inside which all the existing reactors are placed in practice. Table 2.4 presents a description of the hydraulic models used in the representation of stabilisation ponds.

Table 2.4. Characteristics of the hydraulic models more frequently used in the design and performance evaluation of stabilisation ponds

Hydraulic model	Reactor scheme	Characteristics
Plug flow		The fluid particles enter the tank continuously in one end, pass through the reactor, and are then discharged at the other end, in the same sequence in which they entered the reactor. The fluid particles move as a plug, without any longitudinal mixing. The particles maintain their identity and stay in the tank for a period equal to the theoretical hydraulic detention time. This type of flow is reproduced in long tanks with a large length-to-breadth ratio, in which longitudinal dispersion is minimal. Plug-flow reactors are idealised reactors, since complete absence of longitudinal dispersion is difficult to obtain in practice.
Complete mix		The particles that enter the tank are immediately dispersed in all the reactor body. The influent and effluent flows are continuous. The fluid particles leave the tank in proportion to their statistical population. Complete mixing can be obtained in tanks in which the contents are continuously and uniformly distributed. Complete-mix reactors are also known as CSTR or CFSTR (continuous-flow stirred tank reactors). Complete-mix reactors are idealised reactors, since total and identical dispersion is difficult to obtain in practice.
Complete-mix reactor in series		Complete-mix reactors in series are used to model the hydraulic regime of ponds in series or the regime that exists between the idealised plug flow and complete mix. If the series is composed of only one reactor, the system reproduces a complete-mix reactor. If the system has an infinite number of reactors in series, plug flow is reproduced. Influent and effluent flows are continuous. Reactors in series are also commonly applied to maturation ponds.
Dispersed flow		Dispersed or arbitrary flow is obtained in any reactor with an intermediate degree of mixing between the two idealised extremes of plug flow and complete mix. In reality, most reactors present dispersed-flow conditions. However, because of the greater difficulty in their modelling, the flow pattern is frequently represented by one of the two idealised hydraulic models. The influent and effluent flows are continuous.

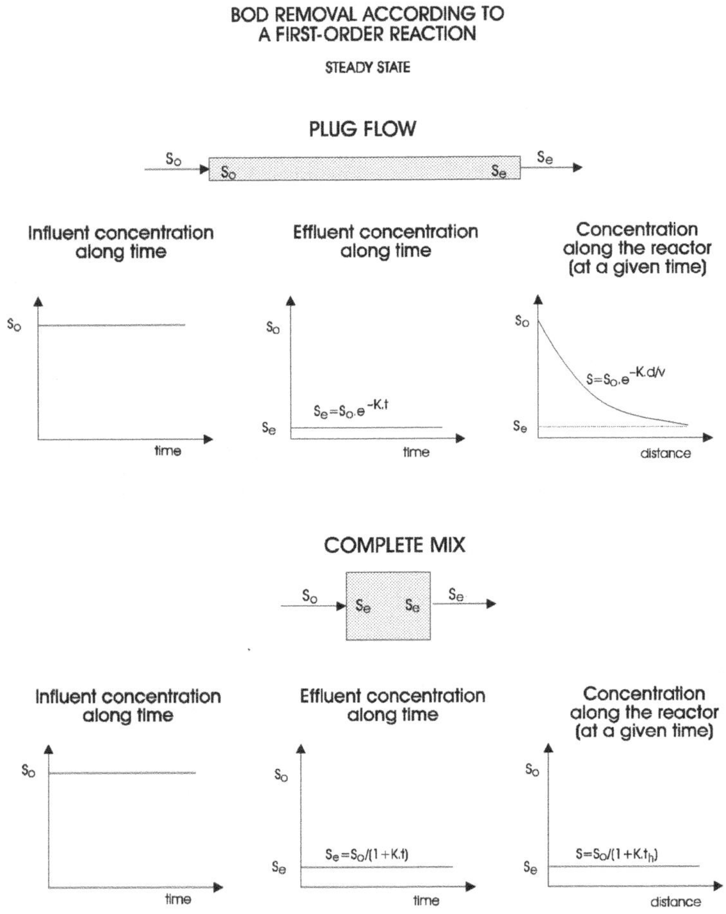

**BOD REMOVAL ACCORDING TO
A FIRST-ORDER REACTION**

STEADY STATE

PLUG FLOW

	Influent concentration along time	Effluent concentration along time	Concentration along the reactor (at a given time)

$S_e = S_o.e^{-K.t}$

$S = S_o.e^{-K.d/v}$

COMPLETE MIX

	Influent concentration along time	Effluent concentration along time	Concentration along the reactor (at a given time)

$S_e = S_o/(1 + K.t)$

$S = S_o/(1 + K.t_h)$

Figure 2.10. BOD removal according to first-order kinetics in plug-flow and complete-mix reactors (S_o = influent total BOD concentration; S = concentration of soluble BOD at a certain distance or time; S_e = effluent soluble BOD concentration; t = operating time; d = horizontal distance along the reactor; v = horizontal velocity; t_h = hydraulic detention time).

The efficiency in the removal of pollutants that are modelled according to first-order reactions (e.g. BOD and coliforms) follows the order presented below:

–	*plug flow pond*	Greater efficiency
–	*series of complete-mix ponds*	⇩
–	*single complete-mix pond*	Lower efficiency

The dispersed-flow regime is not listed above because it can represent well reactors that approach both plug-flow and complete-mix conditions.

Figure 2.10 illustrates the behaviour of the BOD concentration in ponds according to the idealised plug-flow and complete-mix regimes, assuming a first-order removal reaction.

Table 2.5. Formulas for the calculation of the effluent soluble BOD concentration (S)

Hydraulic regime	Scheme	Formula for the soluble effluent BOD_5 concentration
Plug flow		$S = S_o e^{-K.t}$
Complete mix (single cell)		$S = \dfrac{S_o}{1 + K.t}$
Complete mix (equal cells in series)		$S = \dfrac{S_o}{(1 + K\,t/n)^n}$
Dispersed flow		$S = S_o \cdot \dfrac{4ae^{1/2d}}{(1 + a)^2 e^{a/2d} - (1 - a)^2 e^{-a/2d}}$ $a = \sqrt{1 + 4K.t.d}$

S_o = total influent BOD concentration (mg/L)	t = total detention time in the system (d)
S = soluble effluent BOD concentration (mg/L)	n = number of ponds in series (–)
K = BOD removal coefficient (d^{-1})	d = dispersion number (dimensionless)

Table 2.5 presents the formulas for the determination of the soluble effluent BOD concentration for the various hydraulic regimes.

2.6.2 Soluble and particulate effluent BOD

It should be noticed that, in Table 2.5, S is the *soluble* **effluent BOD**. The **influent BOD S_o** is considered to be the *total* **BOD** (soluble + particulate), because the organic suspended solids, responsible for the particulate BOD, are converted into soluble organic matter, through the action of enzymes released into the medium by the bacteria themselves. Therefore, the bacteria assimilate the original soluble BOD of the wastewater (rapid assimilation) and the particulate BOD (after conversion to soluble BOD). Hence, in principle, the total BOD (soluble + particulate) would be available for the bacteria.

The **total effluent BOD** is also associated with two components:

- *soluble BOD*: mostly remaining BOD from the influent wastewater after treatment
- *particulate BOD*: BOD caused by the suspended solids in the effluent

The suspended solids in the effluent of facultative ponds are predominantly algae that may or may not exert some oxygen demand in the receiving water body, depending on their survival conditions. The following comments can be made (Arceivala, 1981; Abdel-Razik, 1991; Mara et al, 1997):

- If the algae die, the stabilisation of the organic fraction of their cellular mass will consume oxygen.

- If the algae are consumed by zooplankton and enter the food web, this can be advantageous for having a more productive environment, useful, for instance, for fish culture.
- If the algae continue to multiply themselves in the receiving water, they can lead to the beneficial effect of oxygen production. The algae undertake photosynthesis as well as respiration, but the amount of oxygen produced by photosynthesis during the sunny hours of the day is much greater than that consumed for respiration during the 24 hours of the day.
- If the effluent is used for irrigation, the algae can also be beneficial. Cyanobacteria contribute to the fixation of nitrogen, and other algae, when dead, slowly release nutrients used by the plants. Besides that, they increase the organic matter in the soil, enhancing its water retention capacity. However, excessive concentrations of algae can affect the soil porosity.

According to Mara (1995), the suspended solids from facultative ponds are about 60 to 90% algae. Each 1 mg of algae generates a BOD_5 around 0.45 mg. Consequently, 1 mg/L of suspended solids in the effluent is capable of generating a BOD_5 (in the BOD test, and not necessarily in the receiving body) in the range of $0.6 \times 0.45 \approx 0.3$ mg/L to $0.9 \times 0.45 \approx 0.4$ mg/L:

$$\boxed{1 \text{ mg SS/L} = 0.3 \text{ to } 0.4 \text{ mgBOD}_5/\text{L}}$$

Monitoring of some ponds in Brazil also leads to the following relationship, expressed in terms of COD:

$$\boxed{1 \text{ mg SS/L} = 1.0 \text{ to } 1.5 \text{ mgCOD/L}}$$

Owing to the uncertainty regarding these aspects, a practical approach can be the one of not considering the BOD from the algae (or from the suspended solids) in the effluent from facultative ponds. As a result, the BOD of the effluent from facultative ponds can be considered as being just the **soluble BOD**. In fact, the European Community established the following standards for the effluents from stabilisation ponds (Council of the European Communities, 1991):

- Soluble (filtered) $BOD_5 \leq 25$ mg/L
- Soluble (filtered COD) ≤ 125 mg/L
- Suspended solids ≤ 150 mg/L

The legislation from most countries makes no distinction between the BOD forms, and considers for the discharge standards the values of total BOD. The SS concentration in the effluent from facultative ponds usually complies with the European Community standards, although there can be occasional periods with values greater than those specified.

Unfortunately, there is no mathematical model that gives a reliable prediction of the suspended solids concentration in the effluent from a facultative pond, because of their great temporal variability as a function of the environmental conditions. For design purposes, the estimation of the particulate BOD may be based on effluent SS in the following range:

$$\text{SS effluent} = 60 \text{ to } 100 \text{ mg/L}$$

2.6.3 BOD removal according to the complete-mix model

Since the length / breadth (L/B) ratio usually employed in primary facultative ponds is in the order from 2 to 4, the hydraulic regime that occurs in fact is the dispersed flow (see Section 2.6.4). However, in the design of facultative ponds the complete-mix model (for one or more ponds) has been more frequently adopted due to the following reasons:

- The calculations with the complete-mix model are simpler.
- Facultative ponds are not especially elongated, and deviations from a complete-mix reactor are not substantial.
- Most of the BOD removal coefficients available in literature are for the complete-mix model.
- There is no need to determine the dispersion number of the pond

The value of the BOD removal coefficient (K) was obtained by several researchers at different existing ponds as a function of the influent and effluent BOD concentrations and the detention time. The value of K is always calculated as a function of the assumed hydraulic model. As a result, the values of K reported in the literature are associated with the hydraulic regime, and this fact needs to be taken into account when selecting the value to be adopted in the design of a new pond. As commented, most of the authors assume the complete-mix regime, but this hypothesis is not always explicit when presenting the values of K. When obtaining the value of K based on experimental data, the temperature, flow and the main geometric relationships of the pond (depth, length and breadth) must always be reported, besides the hydraulic model assumed in the calculations. Another point to remember is that, in the estimation of the K values, the BOD values to be considered are: (a) influent BOD: total BOD; (b) effluent BOD: soluble BOD.

For the most frequent case of the design according to the complete-mix model, the following range of K values may be used for design (Silva and Mara, 1979; Arceivala, 1981; EPA, 1983; von Sperling, 2001):

Pond	K value (20 °C)
Primary ponds (receiving raw wastewater)	0.30 to 0.40 d^{-1}
Secondary ponds (receiving effluent from a previous pond or reactor)	0.25 to 0.32 d^{-1}

Von Sperling (2001), analysing the BOD removal in 10 primary and secondary stabilisation ponds in Brazil, found the following mean values: primary ponds: $K = 0.40$ d^{-1} (4 data); secondary ponds: $K = 0.27$ d^{-1} (6 data); all the ponds: $K = 0.32$ d^{-1} (10 data). The value of K equal to 0.25 d^{-1} (for COD) was found by the author and co-workers in a facultative pond treating effluent from a UASB reactor.

It is natural that the BOD removal coefficient is higher in primary facultative ponds, since the raw wastewater contains more easily biodegradable organic matter. On the other hand, the effluent from anaerobic ponds or anaerobic reactors has a more slowly biodegradable organic matter, since the more easily degradable fraction has been already removed in them. Consequently, the secondary facultative ponds, maturation ponds or polishing ponds should have lower K values.

For different temperatures, the value of K can be corrected using the following equation:

$$\boxed{K_T = K_{20} . \theta^{(T-20)}} \qquad (2.7)$$

where:

K_T = BOD removal coefficient at a temperature T (d^{-1})
K_{20} = BOD removal coefficient at a temperature of 20°C (d^{-1})
T = liquid temperature (°C)
θ = temperature coefficient (−)

It should be noted that different values of θ are proposed in the literature. For $K = 0.35$ d^{-1}, mentioned by EPA (1983), the temperature coefficient is $\theta = \mathbf{1.085}$. For $K = 0.30$ d^{-1}, mentioned by Silva and Mara (1979), the reported value is $\theta = \mathbf{1.05}$.

When designing ponds or wastewater treatment plants, one should always keep in mind that the uncertainty in the design is not just in the coefficients of the model, but also in all the input data, starting from the design population and inflow. The design should always have this uncertainty in perspective, in order not to exaggerate in the sophistication in obtaining some coefficients, and forgetting to analyse the reliability of other data, which are possibly more influential (von Sperling, 1995a).

Example 2.1 illustrates the determination of the effluent soluble BOD concentration and the calculation of the resulting removal efficiency, for a given detention time and an adopted K value.

Example 2.1

Calculate the effluent soluble BOD concentration (S) in the following facultative ponds systems: (a) one plug-flow cell; (b) two complete-mix cells in series; (c) one complete-mix cell. Data:

- influent total BOD: $S_0 = 300$ mg/L
- BOD removal coefficient: $K = 0.30$ d^{-1} (adopted, for all the systems)

Example 2.1 (Continued)

- total detention time: $t = 30$ days
- liquid temperature: $20\ ^\circ$C

Solution:

Using the formulas from Table 2.5:

Hydraulic model	Formula		Soluble BOD (S) (mg/L)	Efficiency E (%)
Ideal plug flow (1 cell)	$S = S_o e^{-K.t}$	$S = 300.e^{-0,30\times30}$	≤ 1	99.99
Ideal complete mix (2 cells)	$S = \dfrac{S_o}{(1+K\frac{t}{n})^n}$	$S = \dfrac{300}{(1+0.30\times\frac{30}{2})^2}$	10	97
Ideal complete mix (1 cell)	$S = \dfrac{S_o}{1+K.t}$	$S = \dfrac{300}{1+0.30\times30}$	30	90

Efficiency: $E = (S_o - S).100/S_o$

Comments:

- Greatest efficiency is obtained with the plug-flow reactor
- Cells in series are more efficient than a single cell
- The results are obtained assuming that the ponds behave as ideal reactors, and that the value of K is the same, independent of the hydraulic regime
- For primary facultative ponds, the plug-flow model is not adequate, since the geometry of the ponds is not of a very elongated rectangle, in order to avoid organic overloading close to the inlet zone of the pond
- The calculated efficiencies are based on the soluble BOD in the effluent, and do not take into account the particulate BOD, also present in the ponds effluent.

2.6.4 BOD removal according to the dispersed-flow model

In reality, the hydraulic regime in a stabilisation pond does not exactly follow the ideal complete-mix or plug-flow models, but an intermediate model. The complete-mix and plug-flow models constitute an envelope, inside which all the reactors in reality are located. The **complete-mix** model represents one extreme (*infinite longitudinal dispersion*), while the **plug-flow** model represents the other extreme (*no longitudinal dispersion*). Inside these extremes are located the reactors modelled according to the **dispersed flow**, comprising all the ponds found in practice. For this reason, the knowledge of the dispersed-flow model is important, since it can be used as a better approximation for the design of stabilisation ponds.

However, modelling of a pond according to the dispersed flow model is more complicated, due to the need of two parameters (BOD removal coefficient and dispersion number), unlike the previous models, in which the knowledge of only the BOD removal coefficient is needed.

a) BOD removal coefficient K

The value of the BOD removal coefficient (K) can be obtained through one of the following empirical relations, obtained in studies of ponds modelled according to the dispersed flow regime:

- Arceivala (1981), after some simplifications by the author:

$$K = 0.132 \times (\log_{10} L_s) - 0.146 \qquad (2.8)$$

- Vidal (1983), after some simplifications by the author:

$$K = 0.091 + 2.05 \times 10^{-4}.L_s \qquad (2.9)$$

It should be highlighted that the temperature coefficient (θ) for Arceivala's equation is **1.035**, differently from the coefficients expressed in Item 2.6.3. With relation to Vidal's equation, the temperature correction was not expressed in the usual Arrhenius form, but through analysis of the original formula, a value of θ lower than 1.035 is obtained.

Table 2.6 presents the values of K according to Arceivala and Vidal for different surface loading rates (for a liquid temperature of 20 °C and inside of the validity range of the equations). It can be observed that the values of K obtained by the two formulas are very similar. Experimental data obtained by the author and co-workers in facultative ponds acting as post-treatment for the effluent of UASB reactors showed good agreement with the removal coefficients K obtained with both equations.

Table 2.6. Values of the BOD removal coefficient (K, in d^{-1}) as a function of the surface loading rate, for the dispersed flow model (20 °C)

Equation	L_s (kgBOD$_5$/ha.d)				
	120	140	160	180	200
Arceivala (1981)	0.128	0.137	0.145	0.152	0.158
Vidal (1983)	0.116	0.120	0.124	0.128	0.132

b) Dispersion Number d

The other parameter to be determined is the Dispersion Number (d), which is expressed by Equation 2.10.

$$d = D/U.L = D.t/L^2 \qquad (2.10)$$

where:
  d = Dispersion Number (−)
  D = longitudinal dispersion coefficient (m^2/d)

U = mean longitudinal velocity along the reactor (m/d)
L = longitudinal length along the reactor (m)

When **d** tends to *infinity*, the reactor tends to the **complete-mix** regime. On the other hand, when **d** tends to *zero*, the reactor tends to the **plug-flow** regime.

The dispersion coefficient D is needed for the calculation of **d**. In existing reactors, D can be obtained experimentally by means of tests with tracers. For the design of new ponds, **d** is of course unknown, and its future value should be estimated according to some criterion. The literature presents some empirical relationships that can be used for this preliminary estimation:

- Polprasert and Batharai (1983):

$$d = \frac{0.184.t.\nu.(B + 2.H)^{0.489}.B^{1.511}}{(L.H)^{1.489}} \qquad (2.11)$$

- Agunwamba et al (1992), original formula simplified by the author:

$$d = 0.102.\left(\frac{3.(B + 2.H).t.\nu}{4.L.B.H}\right)^{-0.410}.\left(\frac{H}{L}\right).\left(\frac{H}{B}\right)^{-(0.981 + 1.385.H/B)} \qquad (2.12)$$

- Yanez (1993)

$$d = \frac{(L/B)}{-0.261 + 0.254.(L/B) + 1.014.(L/B)^2} \qquad (2.13)$$

- Von Sperling (1999)

$$d = \frac{1}{(L/B)} \qquad (2.14)$$

where:
   L = length of the pond (m)
   B = breadth of the pond (m)
   H = depth of the pond (m)
   t = detention time (d)
   ν = kinematic viscosity of the water (m^2/d)

The kinematic viscosity of the water is a function of the temperature (see Table 2.7). Based on the data from Table 2.7, von Sperling (1999) proposed a correlation for the kinematic viscosity of the water as a function of the temperature (Equation 2.15).

$$\nu = 0.325.T^{-0.450} \qquad (2.15)$$

(for T = 10 to 30 °C, R^2 = 0.986)

Table 2.7. Relation between the kinematic viscosity and the
temperature of the water

Water temperature (°C)	Kinematic viscosity (m²/d)
10	0.113
15	0.098
20	0.087
25	0.077
30	0.069

Source: Metcalf & Eddy (1991)

Table 2.8. Ranges of values of the Dispersion Number d, obtained through the use of
the Agunwamba et al (1992), Yanez (1993) and von Sperling (1999) equations

Model	Length (m)	Depth (m)	L/B = 1	L/B = 2 to 4	L/B = 5 to 10
Agunwamba (Eq. 2.12)	L ≤ 100	1.5	0.4–0.7	0.1–0.4	0.03–0.17
		2.5	0.5–0.9	0.1–0.5	0.02–0.22
	L > 100	1.5	0.6–1.1	0.2–0.5	0.07–0.23
		2.5	0.7–1.3	0.2–0.7	0.10–0.30
Yanez (Eq. 2.13)	–	–	1.0	0.24–0.46	0.1–0.2
von Sperling (Eq. 2.14)	–	–	1.0	0.25–0.5	0.1–0.2

Limits for the utilisation of Agunwamba's equation in this table: t = 20 to 40 d; L ≤ 300 m;
T = 20 °C
In each column, for each range of L/B ratios, the smallest value of d corresponds to the largest
L/B value

It should be highlighted that the dispersion number d can vary with time, in the
same pond, as a result of the variation of environmental conditions, which affect
the hydrodynamics of the pond. Kellner and Pires (1998) emphasise the limitations
associated to the estimation of the dispersion in the pond, which should always be
present in the interpretation of operational results.

However, in terms of design, a practical approach is needed, leading to the use
of the empirical formulas. Equation 2.12 (Agunwamba et al, 1992) was reported
to give a better fit to the experimental data than Equation 2.11 (Polprasert and
Agarwalla, 1994). Table 2.8 presents ranges of average values of d obtained using
Equations 2.12, 2.13 and 2.14. The equations of Agunwamba and Yanez lead to
similar results, for ponds with lengths greater than 100 m. The equation of von
Sperling is essentially a simplification of the Yanez equation, leading to practically
the same values.

An additional comparison between the four Dispersion Number estimation
methods was done by von Sperling (2003). A series of 1000 randomly generated
independent sets of physical data was used to compare the values of d resulting
from the four methods. In each one of the 1000 groups, the input data varied randomly, covering most of the situations found in practice. The ranges of variation

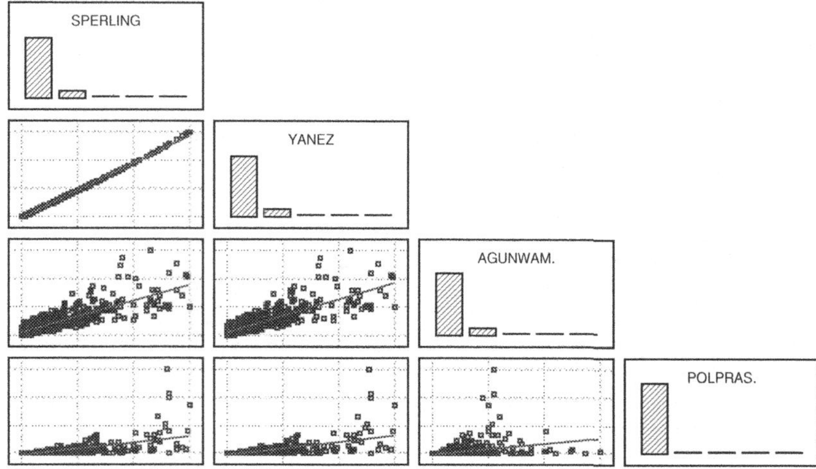

Figure 2.11.   Scatter-plot of the 1000 values of d, generated according to the von Sperling, Yanez, Agunwamba et al and Polprasert and Batharai models

were: (a) length/breadth ratio: L/B = 1 to 16; (b) length of the pond: L = 20 to 300 m; (c) depth of the pond: H = 1.0 to 3.0 m; (d) hydraulic detention time: t = 3 to 40 d; (e) liquid temperature: T = 15 to 25 °C.

Figure 2.11 shows the scatter-plot of the 1000 results of the Dispersion Number d obtained, according to the four methods. From the figure, it is clearly observed that: (a) the von Sperling and Yanez models lead to practically the same results, in all the d range values; (b) the Agunwamba model produces results close to the von Sperling and Yanez models, especially for lower d values; (c) the Polprasert model generates values that are very different from the three other models, especially in the upper half of the d values (according to von Sperling and Yanez), and in all the d values range (according to the Agunwamba model).

c)   Relationship between removal coefficients for different hydraulic regimes

With respect to the removal coefficient K, in principle, its value should be the same for the complete-mix regime as well as for the plug-flow regime. However, in existing ponds, in most cases the value of K is estimated assuming the complete-mix model, knowing the BOD concentrations at the inlet (S_o) and outlet (S) and the detention time (t). Through rearrangement of the equation for S for the complete-mix model (Table 2.5), the value of K can be obtained. In this case, the value of K is overestimated, because in reality, the hydraulic regime is not the ideal complete mix, but the dispersed flow. Even for a square pond, the dispersion number d is equal to 1.0 (according to Yanez and von Sperling – Equations 2.13

Table 2.9. Ratio between the removal coefficients K obtained in the complete-mix regime
and in the plug-flow regime, for different values of K.t (dispersed flow) and
of the dispersion number d

K.t (dispersed flow)	K (complete mix) / K (dispersed flow)			
	$d = 1.0$ $L/B \approx 1$	$d = 0.5$ $L/B \approx 2$	$d = 0.2$ $L/B \approx 4$	$d = 0.1$ $L/B \approx 10$
0	1.00	1.00	1.00	1.00
1	1.14	1.23	1.40	1.52
2	1.29	1.52	1.95	2.32
3	1.46	1.83	2.68	3.55
4	1.64	2.21	3.66	5.39
5	1.83	2.65	4.95	8.18
6	2.04	3.15	6.62	12.28
7	2.27	3.73	8.81	18.21
8	2.53	4.39	11.60	26.81
9	2.79	5.14	15.16	39.11
10	3.08	6.01	19.66	56.50

and 2.14), which is far away from the higher values that characterise ideal
complete mix.

Table 2.9 presents the correspondence between the values of K calculated according to the two hydraulic regimes (complete mix and dispersed flow), for different values of d (or L/B ratio) and the dimensionless pair K.t (for dispersed
flow). For example, in a pond with an L/B ratio $\approx$ 2 (d = 0.5), detention time
t = 27 d, K (dispersed flow) = **0.15 d^{-1}**, one has: K.t = 27 × 0.15 $\approx$4. The ratio
K (complete mix) / K (dispersed flow) is, according to Table 2.9, for K.t = 4 and
d = 0.5, equal to 2.21. This means that, if the coefficient K were determined in
this pond assuming complete mix, a value of K = 2.21 × 0.15 = **0.33 d^{-1}** would
be obtained. The values of K mentioned in the literature for the complete-mix
regime are between 0.25 and 0.40 d^{-1} (see Section 2.6.3), that is, close to the value
of 0.33 d^{-1} obtained in this example. However, if the pond had other geometric
relationships and other detention times, the conversion of the coefficients could
lead to very different values.

d) Removal efficiency

With the values of d and K (dispersed flow), the efficiency of the pond in the
removal of BOD can be estimated, using the formulas presented in Table 2.5 for
dispersed-flow reactors. In these equations, when d = 0, the formula produces results practically equal to those of the plug-flow equation. Similarly, when d = ∞ or,
in practical terms, very high, the results are very close to those of the complete-mix
equation. In the dispersed-flow formula, the second term of the denominator can
be ignored, because it is usually very small. Example 2.2 illustrates the calculation
for a pond of conventional dimensional relations.

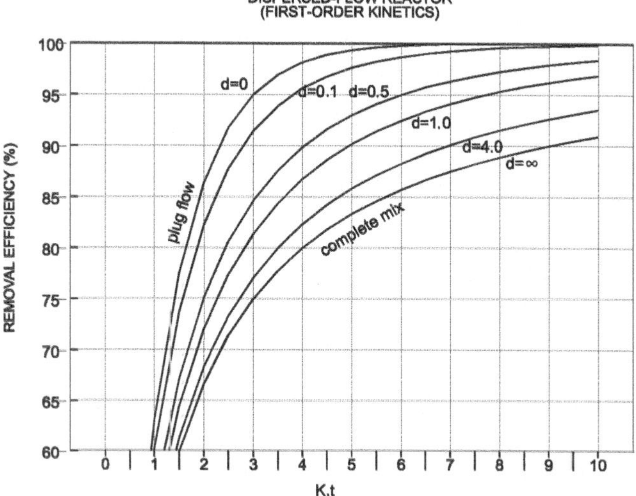

Figure 2.12.   Removal efficiency of a compound following a first-order reaction (e.g.: BOD), for the main hydraulic models

To visualise these concepts, Figure 2.12 plots the dimensionless product K.t versus the BOD removal efficiency, based on a rearrangement of the classical graph of Thirumurty (1969).

Example 2.2

Calculate the effluent soluble BOD concentration (S) according to the dispersed-flow model, for a pond with the following data:

- Influent total BOD: $S_0 = 300$ mg/L
- BOD removal coefficient for the dispersed-flow model: $K = 0.15$ d^{-1} (adopted, see Table 2.5)
- Detention time: $t = 30$ d
- Ratio length/breadth: $L/B = 2$

Solution:

a)   Estimation of the dispersion number d

Considering that the L/B ratio is equal to 2, the dispersion number d is between 0.4 and 0.5, according to the formulas of Agunwamba et al, Yanez and von Sperling (see equations 2.12 to 2.14). Adopt, in the present example, the value of 0.4.

Example 2.2 (Continued)

b)   Calculation of the effluent concentration S

According to the formulas presented in Table 2.5 for the dispersed-flow model:

$$a = \sqrt{1 + 4K.t.d} = \sqrt{1 + 4 \times 0.15 \times 30 \times 0.4} = 2.86$$

$$S = S_o.\frac{4ae^{1/2d}}{(1 + a)^2 e^{a/2d} - (1 - a)^2 e^{-a/2d}}$$

$$= 300.\frac{4 \times 2.86.e^{1/(2\times0.4)}}{(1 + 2.86)^2.e^{2.86/(2\times0.4)} - (1 - 2.86)^2.e^{-2.86/(2\times0.4)}} = 22\,\text{mg/L}$$

It can be observed that this value is in-between those obtained in Example 2.1 for one complete-mix cell (S = 10 mg/L) and two complete-mix cells in series (S = 30 mg/L).

The same considerations made above regarding the conversion of the coefficients K (dispersed flow) to K (complete mix) could have been made.

c)   Calculation of the BOD removal efficiency

$$E = 100.(S_o - S)/S_o = 100 \times (300 - 22)/300 = 93\%$$

The same value could have been obtained through Figure 2.12, for d = 0.4 and K.t = 0.15 × 30 = 4.5.

The calculated efficiencies are based on the soluble BOD in the effluent, and do not take into account the particulate BOD, also present in the pond effluent.

2.7 POND ARRANGEMENTS

The system of facultative ponds can be designed to have more than one pond, leading to a higher operational flexibility. When analysing the division of a pond into a larger number of units, the following aspects should be taken into consideration:

- **Cells in series**. In principle, a system of ponds in series, with a certain total detention time, has a greater efficiency than a single pond, with the same total detention time. The implication is that, for the same effluent quality, a smaller area can be occupied with a system of ponds in series. However, organic overloading in the first facultative pond in the series should be considered (see below).
- **Cells in parallel**. A system of ponds in parallel has approximately the same efficiency as a single pond (some difference may occur because of different dispersion numbers between the single pond and each pond in parallel). However, the system has more flexibility and guarantee, in case there is the need to interrupt the flow to a pond, owing to some problem or occasional maintenance (although this should be rare). As a consequence, the operation of the system will not be interrupted.

- **Organic overload in the first cell**. If the ponds are in series, it should be taken into account that the first cell may be overloaded, because it receives the entire influent load, with the possibility of having anaerobic conditions. The design should evaluate the oxygen balance in the first cell (production and consumption), or verify that the surface loading rate is not excessive in the first cell. To minimise this situation, cells of different sizes can be adopted, with the first unit having the largest area. However, the subsequent units could be considered to be more maturation ponds than facultative ponds as such. This overloading aspect is very important in primary ponds (that receive raw sewage), and *frequently restricts the utilisation of facultative ponds in series*. Ponds in series are more used for the removal of pathogens (maturation ponds), in which there should be no problems with organic overloading in the first cell.
- **Internal divisions**. The subdivision of a single pond into a larger number of ponds implies the need of intermediate embankments.
- **Plug flow**. Theoretically, an infinite number of cells in series corresponds to a plug flow, which would be the most efficient system for the removal of BOD. Thus, instead of having a high number of ponds in series, a single pond with a predominantly longitudinal pathway can be adopted, which can be obtained through a series of U-curves or baffles. In this case, the mentioned aspects of organic overloading close to the inlet zone should be taken into consideration. The plug flow is more used for the polishing of the effluent, such as in maturation ponds, in which there is no concern with organic overload in the inlet zone. For facultative ponds, Yanez (1993) suggests a maximum length/breadth ratio of 8:1. However, it is believed that lower ratios, of the order of 2 to 4 can be safer, from the point of view of organic overloading.

2.8 SLUDGE ACCUMULATION

The sludge accumulated in the bottom of the pond is a result of the suspended solids from the influent, including sand, plus settled microorganisms (bacteria and algae). The organic fraction of the sludge is digested anaerobically, being transformed into gases. Hence, the accumulated volume is lower than the settled volume.

The average sludge accumulation rate in facultative ponds is in the order of only 0.03 to 0.08 m^3/inhab.year (Arceivala, 1981). Silva (1993) and Gonçalves (1999) present average sludge thickness values around 1 to 3 cm/year. Because of this low accumulation rate, the occupation of the pond volume is very slow. Unless the pond receives a very high load, the sludge will accumulate for several years without the need for its removal.

From the accumulated sludge, only a small fraction is represented by grit. In spite of this, it can be necessary to remove the accumulated grit, since it tends to concentrate close to the inlet and in the first cell of a system in series. This emphasises the need for good preliminary treatment of the wastewater.

Table 2.10. Connection between the colour of the pond and operational characteristics

Pond colour	Interpretation
Dark green and partially transparent	• Unimportant presence of other microorganisms in the effluent • High pH and DO values • Pond in good conditions
Yellow green or excessively clear	• Growth of rotifers, protozoa or crustaceans, which feed on the algae and can cause their destruction in few days • If the conditions persist, there will be a decrease in DO and an occasional bad smell
Greyish	• Overload of organic matter and/or short detention time • Incomplete fermentation in the sludge layer • The pond should be put out of operation
Milky green	• The pond is in a self-flocculation process as a result of high pH and temperature • Precipitation of manganese and calcium hydroxides, sweeping the algae and other microorganisms
Blue greenish	• Excessive proliferation of cyanobacteria • The bloom of certain species forms a scum that decomposes easily, leading to the release of bad smells, reduction of light penetration and, as a consequence, reduction of oxygen production
Brownish red	• Overload of organic matter • Presence of photosynthetic sulphide-oxidising bacteria (they require light and sulphides, use CO_2 as an electron acceptor, do not produce oxygen and do not help in BOD removal)

Source: Arceivala (1981); CETESB (1989)

The anaerobic digestion of the bottom sludge can generate soluble non-stabilised by-products, which, when reintroduced into the upper liquid mass, are responsible for a new BOD load. This happens at a higher rate in the warmer periods. Thus, the summer months cannot necessarily be the best performance months of the pond (Abdel-Razik, 1991). The impact of this phenomenon will be larger or smaller, depending on the magnitude of the reintroduced BOD load, compared to the influent BOD load.

2.9 OPERATIONAL CHARACTERISTICS

The interpretation of the predominant colour in the pond can reveal important operational conditions (see Table 2.10). Some of these aspects are reviewed in Section 21, relative to maintenance and operation.

2.10 POLISHING OF POND EFFLUENTS

There are several possibilities for improving the quality of pond effluents, mainly aiming at the removal of suspended solids (algae). Some of the technologies are: (a) intermittent sand filters, (b) rock filters, (c) microsieves, (d) ponds with floating macrophytes, (e) land application, (f) wetlands, (g) coagulation and clarification processes, (h) flotation and (i) aerated biofilters (EPA, 1983; WPCF, 1990; Mara et al, 1992; Oliveira and Gonçalves, 1995; Gonçalves et al, 2000; Crites and Tchobanoglous, 2000).

The inclusion of any of these processes should naturally find a justification from the point of view of the needs of the receiving body (and not only as a safeguard in terms of compliance to discharge standards), since they imply an elevation of the treatment costs and complexity.

The post-treatment systems are more applicable for the improvement of the effluent from already existing ponds. Possibly, in new projects, if a high quality effluent in terms of BOD/COD and nutrients is required, other more efficient treatment systems should be adopted from the beginning, instead of the combination of facultative ponds with post-treatment. Some processes for the removal of algae are discussed below.

Rock filters. Rock filters consist of submerged stone porous beds, in which the algae settle, as the water flows through the bed. The algae are decomposed, releasing nutrients that are used by the bacteria growing on the surface of the filter. Besides the removal of algae, nitrification can also occur. The performance depends on the loading rate, temperature and size and shape of the stones. Loading rates are in the order of 1.0 m^3 of effluent per m^3 of rock medium per day. The stones have dimensions of about 50 to 200 mm – larger values reduce the surface exposure area, while smaller values can lead to clogging. The height of the bed is around 1.5 to 2.0 m. The pond effluent should be introduced below the surface layer to avoid odour problems. The unit can be located inside the pond. The costs are low and the operation simple, being associated with the periodic removal of the accumulated humus (Mara et al, 1992). The main disadvantages are associated with the possible generation of bad odours and the fact that the net life and the cleaning procedures are not yet totally established (WPCF, 1990; Crites and Tchobanoglous, 2000).

Intermittent sand filters. Intermittent sand filters are somewhat similar to the slow filters, operated in an intermittent way. The effluent is disposed periodically on the surface of the filter bed. The suspended solids and the organic matter are retained in the first 5 to 8 cm. After clogging, the surface sand layer is removed. The bed layer has a thickness of about 0.5 to 1.0 m, with a sand of effective size between 0.2 and 0.3 mm. The hydraulic loading rate is within 0.2 to 0.6 m^3/m^2.d, with the lower values associated with effluents with SS levels greater than 50 mg/L and cold periods (Crites and Tchobanoglous, 2000).

Floating macrophytes. The use of ponds with water hyacinths (*Eichhornia crassipes*) has been the object of considerable controversy. Water hyacinths are autotrophs (plants), meaning that they do not use the organic matter from the sewage. However, their root system allows the development of a biomass capable of stabilising part of the organic matter, besides adsorbing other pollutants, such as metals. The root system also contributes to a larger sedimentation of the suspended solids. Although there is no consensus on this subject, most people involved directly with the operation of these ponds comment that the problems outweigh the benefits.

Water hyacinths grow very fast, and it is necessary to have an infrastructure for their removal compatible with their growth rate, in order to avoid dead plants sinking to the bottom of the pond, where they undergo anaerobic conversion and allow the resolubilisation of the removed pollutants.

A macrophyte of simpler handling, owing to its smaller size, is the duckweed (*Lemna* sp.). The duckweed develops on the surface of the pond, decreasing the light penetration, which reduces the algal growth rate and leads to a more clarified effluent. Ponds with duckweed should be located at the end of the pond series, since their efficiency is lower than a maturation pond in the removal of coliforms, but they generate an effluent with lower SS levels. The duckweeds can be collected and serve as food for fish in other ponds.

Physical–chemical removal. The removal of SS by coagulation/flocculation can be done in a simple way locating the units inside the pond. Gonçalves et al (2000) inserted a mixing tank (t = 1 min), a granular flocculation unit (t = 7 min) and a laminar settling tank (hydraulic loading rate = 70 m^3/m^2.d) inside a facultative pond. The coagulant that produced better results was ferric chloride, with a dosage of 80 mg/L. Good removals of SS (73%), COD (58%) and phosphorus (83%) were reached. The sludge was recirculated to the anaerobic pond and no alteration in the performance of the anaerobic pond was observed.

Example 2.3

Design a treatment system composed of primary facultative ponds based on the following data:

- Population = 20,000 inhabitants
- Influent flow: Q = 3,000 m^3/d
- Influent BOD: S$_o$ = 350 mg/L
- Temperature: T = 23 °C (mean liquid temperature in the coldest month)

Solution:

a) Calculation of the influent BOD$_5$ load

$$\text{load} = \text{concentration} \times \text{flow} = \frac{350\,\text{g/m}^3 . 3000\,\text{m}^3/\text{d}}{1000\,\text{g/kg}} = 1{,}050\,\text{kg/d}$$

b) Adoption of the surface loading rate

$$L_s = 220\,\text{kgBOD}_5/\text{ha.d (adopted} - \text{see Section 2.5.a)}$$

c) Calculation of the required area

$$A = \frac{L}{L_s} = \frac{1{,}050\,\text{kg/d}}{220\,\text{kg/ha.d}} = 4.8\,\text{ha} = 48{,}000\,\text{m}^2$$

d) Adoption of a value for the pond depth

$$H = 1.80\,\text{m (adopted)}$$

Example 2.3 (Continued)

e) Calculation of the resulting volume

$$V = A.H = 48,000\,m^2 \times 1.80\,m = 86,400\,m^3$$

f) Calculation of the resulting detention time

$$t = \frac{V}{Q} = \frac{86,400\,m^3}{3,000\,m^3/d} = 28.8\,d$$

g) Adoption of a value for the BOD removal coefficient (K)

- Complete-mix regime, at 20 °C (see Section 2.6.3):

$$K = 0.35 d^{-1}$$

- Correction for the temperature of 23 °C:

Adopting a value for the temperature coefficient $\theta = 1.05$:

$$K_T = K_{20}.\theta^{(T-20)} = 0.35 \times 1.05^{(23-20)} = 0.41\,d^{-1}$$

h) Estimation of the effluent soluble BOD

Using the complete-mix model (considering a not predominantly longitudinal cell):

$$S = \frac{S_o}{1 + K.t} = \frac{350}{1 + 0.41 \times 28.8} = 27\,mg/L$$

Note: if the dispersed-flow model had been adopted, with the dimensions L, B and H determined in item m below, together with equations from Section 2.6 (Table 2.5, Equations 2.12, 2.13 or 2.14, $K = 0.15\ d^{-1}$ for 20 °C, $\theta = 1.035$), this would have lead to:

- $d = 0.35$ (according to Eq. 2.12) , $d = 0.37$ (according to Eq. 2.13) or $d = 0.40$ (according to Eq. 2.14)
- $S = 23$ mg/L

i) Estimation of the effluent particulate BOD

Assuming an effluent SS concentration equal to 80 mg/L, and considering that each 1 mgSS/L implies a BOD_5 of around 0.35 mg/L (see Section 2.6.2):

Particulate $BOD_5 = 0.35$ mgBOD_5/mgSS $\times$ 80 mgSS/L $= 28$ mgBOD5/L

It should be remembered that the particulate BOD is detected in the BOD test, but it may not be exerted in the receiving body, depending on the survival conditions of the algae.

Example 2.3 (Continued)

j) Total effluent BOD

Total effluent BOD = Soluble BOD + Particulate BOD
Total effluent BOD = 27 + 28 = 55 mg/L

l) Calculation of the BOD removal efficiency

$$E = \frac{S_o - S}{S_o}.100 = \frac{350 - 55}{350}.100 = 84\%$$

m) Dimensions of the pond

The dimensions of the pond are a function of the local area and topography. For the purposes of this example, unspecific values will be adopted.

If two ponds in parallel and a length/breadth (L/B) ratio equal to 2.5 in each pond are adopted, one has:

Area of 1 pond = 48,000/2 = 24,000 m^2

A = L.B = [(L/B).B].B = [2.5.B].B = 2.5.B^2

24,000 m^2 = 2.5.B^2 → B = [A/ (L/B)]$^{0.5}$ = (24,000/2.5)$^{0.5}$ = 98.0 m
L = (L/B) × B = 2.5.B = 2.5 × 98.0 m = 245.0 m

- Length: L = 245.0 m
- Breadth: B = 98.0 m

n) Total area required for the whole system

The total area required for the ponds, including the embankments, urbanisation, internal roads, laboratory, parking and others, is about 25% to 33% greater than the net area calculated at mid-depth (Arceivala, 1981). Hence:

$$A_{total} = 1.3.A_{net} = 1.3 \times 48,000\,m^2 \cong 62,400\,m^2 (6.2\,ha)$$

$$\text{Per capita land requirements} = \frac{62,400\ m^2}{20,000\ hab} = 3.1\ m^2\ /inhab.$$

o) Sludge accumulation

$$\text{Accumulation per year} = 0.05 m^3/inhab. \times 20,000\ inhab. = 1,000\ m^3/year$$

Thickness in 1 year:

$$\text{Thickness} = \frac{1,000\ m^3/year \times 1\ year}{48,000\ m^2} = 0.021\ m/year = 2.1\ cm/year$$

Example 2.3 (Continued)

Thickness in 20 years of operation:
Thickness: 2.1 cm/year × 20 years = 42 cm in 20 years
After 20 years of operation, the sludge occupies only 23% (= 0.42 m/1.80 m) of the liquid depth of the pond.

p) Layout of the system

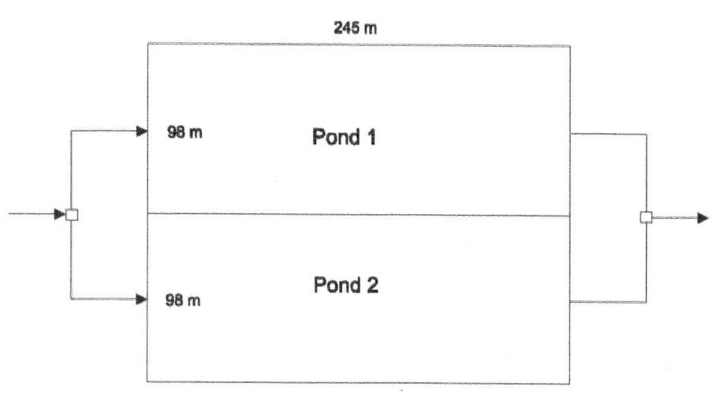

3

System of anaerobic ponds followed by facultative ponds

3.1 INTRODUCTION

Anaerobic ponds constitute an alternative form of treatment, in which the existence of *strictly anaerobic* conditions is essential. This is reached through the application of a high BOD load per unit of volume of the pond, which causes the oxygen consumption rate to be several times greater than the oxygen production rate. In the oxygen balance, the production by photosynthesis and atmospheric reaeration are, in this case, negligible.

Anaerobic ponds have been used for the treatment of domestic sewage and organic industrial wastewaters, with high BOD concentrations, such as slaughterhouses, piggery wastes, dairies, beverage industries, etc.

The conversion of organic matter under anaerobic conditions is slow, owing to the slow growth rate of anaerobic bacteria. This results from the fact that the anaerobic reactions generate less energy than the aerobic reactions for the stabilisation of organic matter. The temperature of the medium has a great influence in the biomass reproduction and substrate conversion rates, which makes warm-climate regions to be favourable for the utilisation of this type of pond.

Anaerobic ponds are usually deep, of the order of 3 m to 5 m. The depth is important, in order to reduce the possibility of the penetration of the oxygen produced in the surface to the other layers. Because these ponds are deeper, the land requirements are correspondingly small.

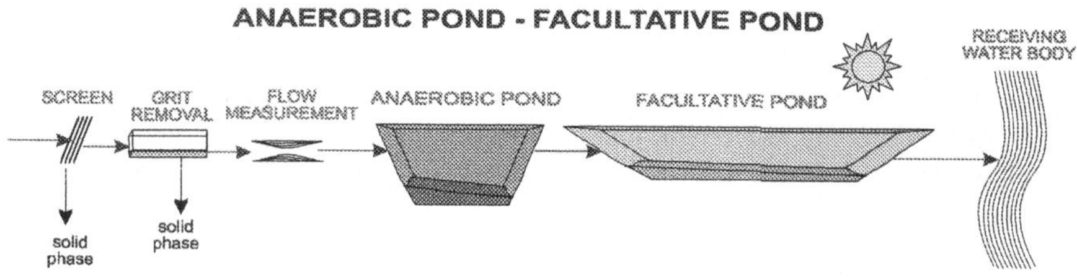

Figure 3.1.   System of anaerobic ponds followed by facultative ponds

Anaerobic ponds do not require any special equipment and have a practically negligible energy consumption (for a possible pumping of the raw sewage or the recirculation of the final effluent).

The BOD removal efficiency in anaerobic ponds is usually of the order of 50% to 70%. The effluent BOD is still high and implies the need of a post-treatment unit. The most widely used post-treatment units are facultative ponds, composing the system of anaerobic ponds followed by facultative ponds (Fig. 3.1).

The removal of BOD in the anaerobic pond provides a substantial saving in the area required for the facultative pond, making the total land requirement (anaerobic + facultative ponds) to be around 45% to 70% of the requirement for a primary facultative pond (receiving raw wastewater).

The existence of an anaerobic stage in an open reactor is always a matter of concern, owing to the possibility of the generation of bad odours. If the system is well balanced, the generation of bad smell should not be important, but occasional operational problems can lead to the release of hydrogen sulphide (H_2S), responsible for obnoxious odours. If the sulphate concentration in the influent is lower than 300 mg/L, the production of sulphide should not be problematic (in anaerobic conditions, sulphate is reduced to sulphide). Additionally, if the pH in the pond is close to neutrality, most of the sulphide will be present in the form of the bisulphide ion (HS^-), which is odourless (Mara et al, 1997). Wastewaters with low pH values (industrial effluents or wastewater originated from a water that is soft, with low alkalinity, high acidity or without pH correction) may induce odour problems. As a result of the points above, the anaerobic-facultative ponds system should be located far away from houses (during all the operational life of the ponds).

3.2 DESCRIPTION OF THE PROCESS

In a simplified way, the anaerobic conversion takes place in two stages:

- liquefaction and formation of acids (through the acid-forming bacteria, or acidogenic bacteria)
- formation of methane (through the methane-forming organisms, or methanogenic archaea)

In the first phase, there is no BOD removal, just the conversion of the organic matter to other forms (simpler molecules and then acids). It is in the second stage that BOD is removed, with the organic matter (acids produced in the first stage) being converted mainly to methane and carbon dioxide. The carbon is removed from the liquid medium by the fact that the methane (CH_4) escapes to the atmosphere.

The methane-forming organisms are very sensitive to the environmental conditions. If their reproduction rate is reduced, there will be the accumulation of the acids formed in the first stage, with the following consequences: (a) interruption of the BOD removal process and (b) generation of bad odours, because the acids are very fetid.

Therefore, it is essential that the appropriate balance between the two communities is guaranteed, ensuring the completion of both stages. For the adequate development of the methane-forming archaea, the following conditions should be met:

- absence of dissolved oxygen (methane-forming archaea are *strict anaerobes* and do not survive in the presence of dissolved oxygen)
- adequate temperature of the liquid (above 15 °C)
- adequate pH (close to or above 7)

The anaerobic activity affects the nature of the solids, in such a way that, in the facultative pond, the solids are less prone to fermentation and flotation, besides decomposing more easily.

3.3 DESIGN CRITERIA FOR ANAEROBIC PONDS

The main design parameters for anaerobic ponds are:

- *Volumetric organic loading rate*
- *Detention time*
- *Depth*
- *Geometry (length / breadth ratio)*

The criterion of the *volumetric organic loading rate* is the most important, and is established as a function of the need of a certain pond volume for the conversion of the applied BOD load. The criterion of the *detention time* is based on the time necessary for the reproduction of the anaerobic bacteria.

a)   Volumetric organic loading rate

The volumetric loading rate L_v, the main design parameter for anaerobic ponds, is a function of the temperature. Warmer locations allow a larger loading rate (smaller pond volume). The consideration of the volumetric load is important, because industrial wastewaters can vary widely in the relationship between flow and BOD concentration (load = flow × concentration). Therefore, only the detention time criterion is insufficient.

Table 3.1. Permissible volumetric loading rates for the design of anaerobic ponds as a function of temperature

Mean air temperature in the coldest month – T (°C)	Permissible volumetric loading rate L_V (kgBOD/m^3.d)
10 to 20	$0.02T - 0.10$
20 to 25	$0.01T + 0.10$
> 25	0.35

Source: adapted from Mara (1997)

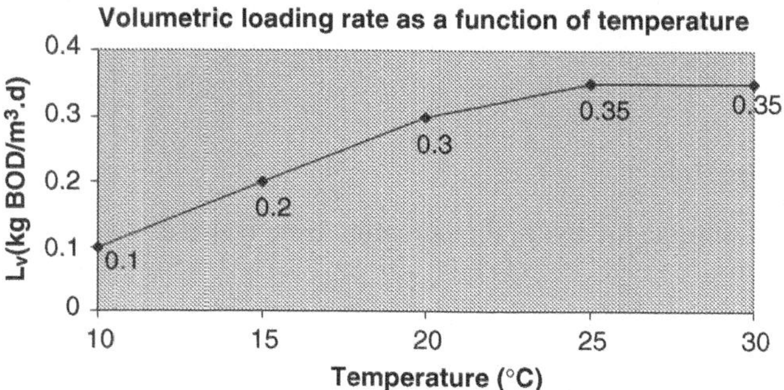

Figure 3.2.   Relation between the permissible volumetric loading rates in anaerobic ponds and the temperature, according to the criteria of Mara (Table 3.1)

Values of volumetric loading rates usually adopted are within the following range:

$$L_v = 0.1 \text{ to } 0.3 \text{ kgBOD}_5/\text{m}^3.\text{d}$$

The upper limit aims at avoiding organic overloading in the anaerobic pond. The lower limit is to avoid that the pond receives a very low organic load, which could give conditions that, under some circumstances, the pond could behave as a facultative pond. This would be harmful to the strictly anaerobic methanogenic archaea.

Mara (1997) proposes the relation between the volumetric loading rates and the temperature presented in Table 3.1 and in Figure 3.2. The values represent maximum permissible rates, and the designer may decide to incorporate more safety by the adoption of lower values of the loading rate.

The volume required is given by:

$$V = L/L_v \qquad (3.1)$$

where:

V = volume required for the pond (m^3)
L = total (soluble + particulate) influent BOD load (kgBOD$_5$/d)
L_v = volumetric loading rate (kgBOD$_5$/m^3.d)

For domestic sewage, the final volume to be adopted for the anaerobic pond is a compromise between the two criteria (detention time and volumetric rate), aiming, as much as possible, to satisfy both. For industrial effluents, the defining criterion is the volumetric loading rate.

In situations in which there is a great variation of the influent load, for example, between the beginning and the end of the design horizon, it is important to verify compliance with the design criteria from the start of the operation. If the initial influent load is low, it may be advisable to divide the implementation into two or more anaerobic ponds in parallel, with only one or some ponds being implemented in the first stage. This assists in guaranteeing that the ponds work under really anaerobic conditions, avoiding very low loading rates.

b)   Detention time

For *domestic sewage*, the hydraulic detention time is usually within the following range:

$$t = 3.0 \text{ d to } 6.0 \text{ d}$$

In conventional anaerobic ponds (in which the inlet pipe is above the sludge layer), if the detention time is lower than 3.0 days, the methane-forming organisms may be washed out of the reactor. In these conditions, the maintenance of a stable bacterial population would not be possible. Apart from the efficiency of the anaerobic pond being reduced, the more serious aspect of imbalance between the acid-forming and methane-forming stages would occur. The consequences would be the accumulation of acids in the liquid, with the generation of bad odours, as a result of the small population of methane-forming organisms to continue the conversion of acids.

However, there is a recent tendency of decreasing the detention times in anaerobic ponds to around **2 days** and, possibly, 1 day. For this, it is necessary to increase the retention time of the biomass and to allow an intimate biomass–wastewater contact. These conditions can be obtained with the distribution of the influent in the bottom of the pond, at several points, aiming at approaching the working principle of an upflow anaerobic sludge blanket reactor. When entering the pond, the influent sewage has direct contact with the anaerobic biomass, optimising the important aspect of the organic matter – biomass contact. Traditional anaerobic ponds that presented operational problems showed an improvement in the performance and a reduction of odour generation with the simple change of the inlet pipe to the bottom of the pond.

With detention times greater than 6 days, the anaerobic pond can behave occasionally as a facultative pond. This is undesirable, because the presence of oxygen is fatal for the methane-forming organisms. *Anaerobic ponds must work as strict anaerobic ponds and cannot alternate between anaerobic, facultative and aerobic conditions.*

After calculating the volume based on the volumetric loading rate (L_v), the resulting detention time is obtained by:

$$\boxed{t = V/Q} \tag{3.2}$$

where:
  t = detention time (d)
  V = volume of the pond (m^3)
  Q = average influent flow (m^3/d)

c) Depth

The depth of anaerobic ponds is high, in order to guarantee the predominance of anaerobic conditions, avoiding the pond to work as a facultative pond. In fact, the deeper the pond, the better. However, deep excavations tend to be more expensive. Values usually adopted are in the range of:

$$\boxed{H = 3.5 \text{ m to } 5.0 \text{ m}}$$

When there is no previous grit removal, the anaerobic pond could have an additional depth of at least 0.5 m, close to the inlet and extending to at least 25% of the area of the pond. However, it is believed that the inclusion of grit chamber units is beneficial, because they minimise problems of grit accumulation close to the inlet pipe and due to their simplicity.

d) Geometry (length / breadth ratio)

Anaerobic ponds are square or slightly rectangular, with typical length/breadth (L/B) ratios of:

$$\boxed{\text{Length / breadth ratio (L/B)} = 1 \text{ to } 3}$$

3.4 ESTIMATION OF THE EFFLUENT BOD CONCENTRATION FROM THE ANAEROBIC POND

There are still no conceptual mathematical models in widespread use that allow an estimation of the effluent BOD concentration from anaerobic ponds. For this reason, these ponds have been designed mainly according to empirical criteria. Mara (1997) proposed the BOD removal efficiencies as a function of the temperature presented in Table 3.2 and illustrated in Figure 3.3.

Table 3.2. BOD removal efficiencies in anaerobic
ponds as a function of the temperature

Mean air temperature of the coldest month - T ($^\circ$C)	BOD removal efficiency E (%)
10 to 25	2T + 20
> 25	70

Source: Mara (1997)

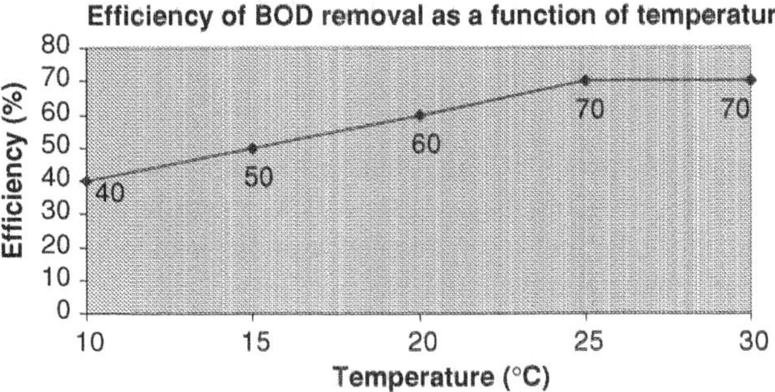

Figure 3.3.   Relationship between the BOD removal efficiency in anaerobic ponds and
the temperature, according to Mara's criterion (Table 3.2)

Once the removal efficiency (E) has been estimated, the effluent concentration
(BOD_{effl}) of the anaerobic pond is calculated using the formulas:

$$E = (S_o - BOD_{effl}).100/S_o \qquad (3.3)$$

or

$$BOD_{effl} = (1 - E/100).S_o \qquad (3.4)$$

where:

S_o = influent total BOD concentration (mg/L)

BOD_{effl} = effluent total BOD concentration (mg/L)

In this empirical approach, the effluent BOD considered is the total BOD,
different from the calculations of facultative ponds, in which the effluent BOD is
split in terms of soluble BOD and particulate BOD.

3.5 DESIGN OF FACULTATIVE PONDS FOLLOWING ANAEROBIC PONDS

Secondary facultative ponds can be designed following the same surface loading rates described in Chapter 2. The resulting detention time will be now smaller, owing to the previous removal of the BOD in the anaerobic pond.

For the design according to the surface loading rate, the BOD concentration and load at the *influent to the facultative pond* are the *effluent from the anaerobic pond*. There are some evidences to suggest that the surface loading rate in secondary facultative ponds could be somewhat higher than those adopted for primary ponds. However, for design purposes, it is better to consider both as being equal for safety reasons (Mara et al, 1992).

In secondary facultative ponds there is more flexibility with regards to the geometry of the pond, which could have higher L/B ratios, since the overloading problems in the inlet zone should be smaller due to the previous removal of a large part of the BOD in the anaerobic pond.

The estimation of the effluent BOD concentration from the facultative pond can be done according to the methodology described in Section 2.6. The removal coefficient K will be in this case lower than in primary facultative ponds, due to the previous removal of the more easily degradable organic matter in the anaerobic pond. The remainder of the organic matter is harder to degrade, implying slower conversion rates. In Section 2.6.3, the following values of K have been suggested for secondary facultative ponds, using the complete-mix model:

$$\boxed{K = 0.25 \text{ to } 0.32 \text{ d}^{-1}}$$

(20 °C, secondary facultative ponds, complete-mix model)

3.6 SLUDGE ACCUMULATION IN ANAEROBIC PONDS

The considerations here are similar to those made in the case of the facultative ponds (Section 2.8). The accumulation rate is in the order of 0.03 to 0.10 m^3/inhab.year (Mendonça, 1990; Gonçalves, 2000), and the lower range is more usual in warm-climate areas. Other data available for accumulation rates are 2 to 8 cm/year (Silva, 1993; CETESB, 1989; Gonçalves, 2000). These values of yearly increases in the thickness of the sludge layer correspond to accumulation rates lower than 0.03 m^3/inhab.year.

The aspects of sludge management in anaerobic ponds are different from facultative ponds. In the latter, the system can operate for several years, eventually during all of the design period, without needing to remove sludge (provided there is a good grit removal in the preliminary treatment). However, because of the smaller volume of the anaerobic ponds, the sludge accumulation manifests itself more rapidly, bringing about the need of an appropriate planning related to the sludge management (see Chapter 11). The anaerobic ponds should be cleaned according to one of the following strategies:

- when the sludge layer reaches approximately 1/3 of the liquid depth
- annual removal of a certain volume, in a pre-determined month, to include the cleaning stage in a systematic way in the operational strategy of the pond

If the removal is not by emptying and drying inside the pond, the whole sludge mass should not be removed, since this would lead to a total loss of the biomass, requiring the anaerobic pond to start up again.

Example 3.1

Design an anaerobic – facultative pond system using the same data from Example 2.3:

- Population = 20,000 inhabitants
- Influent flow = 3,000 m^3/d
- Influent BOD: S_0 = 350 mg/L
- Temperature: T = 23 °C (mean liquid temperature in the coldest month)

Solution:

a) Influent BOD load

From Example 2.3:

$$L = 1,050 \, \text{kgBOD}_5/\text{d}$$

Design of the anaerobic pond

b) Adoption of a value for the volumetric loading rate L_v

$$L_v = 0.15 \, \text{kgBOD/m}^3.\text{d}$$

This is a conservative value (see Section 3.3.a). However, higher values would lead to a smaller pond volume and, as a result, to low detention times (see section d below).

c) Calculation of the required volume

$$\text{volume} = \frac{\text{load}}{\text{volumetric load}} \longrightarrow V = \frac{L}{L_v} = \frac{1,050 \text{ kgBOD/d}}{0.15 \text{ kgBOD/m}^3.\text{d}} = 7,000 \, \text{m}^3$$

d) Verification of the detention time

$$t = \frac{V}{Q} = \frac{7,000 \text{ m}^3}{3,000 \text{ m}^3/\text{d}} = 2.3 \text{ d} \quad \text{OK!}$$

Ponds with such a low detention time should have the inlet at the bottom, in contact with the settled sludge.

Example 3.1 (Continued)

e)   Determination of the required area and dimensions

Depth H = 4.5 m (adopted)

$$\text{area} = \frac{\text{volume}}{\text{depth}} \longrightarrow A = \frac{V}{H} = \frac{7{,}000\,\text{m}^3}{4.5\,\text{m}} = 1{,}556\,\text{m}^2$$

Adopt 2 ponds
Area of each pond: $1{,}556\ \text{m}^2/2 = 778\,\text{m}^2$
Possible dimensions of each pond: 34 m × 23 m

f)   Concentration of effluent BOD

BOD removal efficiency: E = 60% (see Section 3.4)

$$\text{BOD}_{\text{effl}} = (1 - E/100).S_o = (1 - 60/100) \times 350 = 0.4 \times 350 = 140\,\text{mg/L}$$

The effluent from the anaerobic pond is the influent to the facultative pond.

g)   Sludge accumulation in the anaerobic pond

Adopting an accumulation rate of 0.04 m^3/inhab.year (see Section 3.6):

Annual accumulation = 0.04 m^3/inhab.year × 20,000 inhab = 800 m^3/year

Thickness of the sludge layer in 1 year:

$$\text{thickness} = \frac{\text{Annual accumulation} \times \text{time}}{\text{pond area}} = \frac{800\ \text{m}^3/\text{year} \times 1\ \text{year}}{1{,}556\ \text{m}^2}$$

$$= 0.51\ \text{m/year} = 51\,\text{cm/year}$$

This annual accumulation rate, expressed in cm/year, is greater than the values mentioned in Section 3.6, probably because the pond in the present example is deep and has a small detention time (smaller surface area for the sludge to spread itself).

Time to reach 1/3 of the pond depth:

$$\text{time} = \frac{H/3}{\text{yearly thickness}} = \frac{4.5\ \text{m}\ /\ 3}{0.51\ \text{m}\ /\ \text{year}} \cong 2.9\ \text{year}$$

The sludge volume accumulated during this period corresponds to 1/3 of the net pond volume, that is, $7{,}000\ \text{m}^3/3 = 2{,}333\ \text{m}^3$ of sludge.

The sludge should be removed approximately every 3 years (volume of $2{,}333\ \text{m}^3$) or, annually (removal of 800 m^3).

Design of the facultative pond

h)   Influent load to the facultative pond

The effluent load from the anaerobic pond is the influent load to the facultative pond. With the removal efficiency of 60% in the anaerobic pond, the

Example 3.1 (Continued)

influent load to the facultative pond is:

$$L = \frac{(100 - E).L_o}{100} = \frac{(100 - 60) \times 1{,}050}{100} = 420 \, \text{kg BOD/d}$$

i)   Adoption of the surface loading rate

$L_s = 220$ kgBOD/ha.d (equal to the value adopted in Example 2.3)

j)   Required area

$$A = \frac{L}{L_s} = \frac{420 \, \text{kgBOD/d}}{220 \, \text{kgBOD/ha.d}} = 1.9 \, \text{ha} \; (19{,}000 \, \text{m}^2)$$

Adopt two ponds
Area of each pond: 19,000 m^2/2 = 9,500 m^2
Possible dimensions of each pond: L = 155 m and B = 62 m (L/B ratio = 2.5)

k)   Adoption of a value for the depth

H = 1.80 m (adopted)

l)   Calculation of the resulting volume

V = A.H = 19,000 m^2 $\times$ 1.80 m = 34,200 m^3

m)   Calculation of the resulting detention time

$$t = \frac{V}{Q} = \frac{34{,}200 \, \text{m}^3}{3{,}000 \, \text{m}^3/\text{d}} = 11.4 \, \text{d}$$

n)   Adoption of a value for the BOD removal coefficient (K)

-   Complete-mix regime, at 20°C: K = 0.27 d^{-1} (adopted – see Section 3.5)
-   Correction for the temperature of 23°C:

$K_T = K_{20}.\theta^{(T-20)} = 0.25 \times 1.05^{(23-20)} = 0.31 \, \text{d}^{-1}$

o)   Estimation of the effluent soluble BOD

Using the complete-mix model, since the pond is not predominantly longitudinal (length/breadth ratio of 2.5):

$$S = \frac{S_o}{1 + K.t} = \frac{140}{1 + 0.31 \times 11.4} = 31 \, \text{mg/L}$$

p)   Estimation of the effluent particulate BOD

Assuming an effluent SS concentration equal to 80 mg/L, and considering that each 1 mgSS/L leads to a BOD$_5$ of around 0.35 mg/L (see Section 2.6.2):

particulate BOD$_5$ = 0.35 mgBOD$_5$/mgSS $\times$ 80 mgBOD$_5$/L = 28 mgBOD$_5$/L

Example 3.1 (Continued)

It should be remembered that the particulate BOD is detected in the BOD test, but it may not be exerted in the receiving body, depending on the survival conditions of the algae.

q)   Total effluent BOD

total effluent BOD = soluble BOD + particulate BOD

Total effluent BOD = 31 + 28 = 59 mg/L

r)   Calculation of the total BOD removal efficiency of the anaerobic–facultative pond system

$$E = \frac{(S_o - BOD_{eff})}{S_o}.100 = \frac{350 - 59}{350} \times 100 = 83\%$$

s)   Total net area (anaerobic + facultative pond)

$$Total\ net\ area = 0.16\ ha + 1.9\ ha = 2.1\ ha$$

t)   Total area required

The total area is in the order of 25% to 33% greater than the required net area. Thus, the total area occupied by the system of ponds and auxiliary structures is approximately:

$$Total\ area = 1.3 \times 2.1 = 2.7\ ha$$

With primary facultative ponds (Example 2.3), the total area required is 6.2 ha. Therefore, there is a substantial economy of area (56%). The total detention time in the present example is 2.7 d (= 2.3 + 11.4), much lower than that for a primary facultative pond (28.8 m).

It should be remembered that these land requirements are applicable to the current example, which is associated to a relatively high temperature of the liquid, which allows high loading rates and removal efficiencies. In applications in colder places, the required area will be naturally larger.

u)   Layout of the system

ANAEROBIC PONDS - FACULTATIVE PONDS

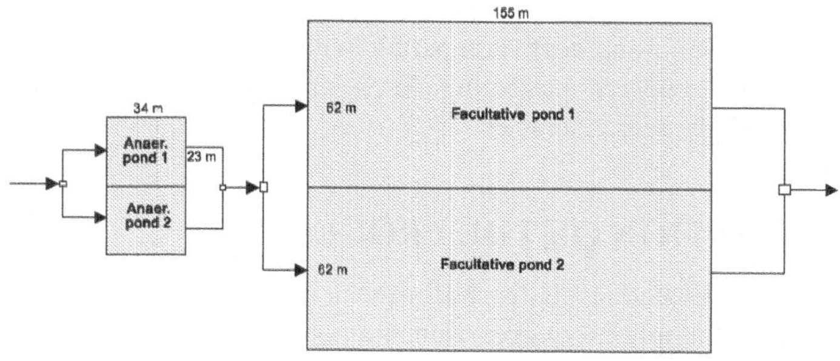

4

Facultative aerated lagoons

4.1 INTRODUCTION

Facultative aerated lagoons (Figure 4.1) are used when it is desired to have a predominantly aerobic system, more compact than facultative ponds or anaerobic-facultative ponds. The main difference with relation to the conventional facultative pond regards the form of oxygen supply. While in facultative ponds the oxygen is obtained from algal photosynthesis, in the case of facultative aerated lagoons the oxygen is supplied by *aerators*.

Because of the introduction of mechanisation, aerated lagoons are less simple in terms of maintenance and operation, compared with conventional facultative ponds. The reduction of the land requirements is therefore obtained with a certain increase in the operational level, besides the introduction of energy consumption.

Overloaded conventional facultative ponds without area for expansion can be converted to facultative aerated lagoons by the inclusion of aerators. However, it is interesting to foresee this possibility in the design period itself, as part of the staging of the plant, so that a pond depth compatible with the future aeration equipment is selected and that concrete protecting plates can be placed at the bottom of the pond, underneath the future aerators.

4.2 DESCRIPTION OF THE PROCESS

The pond is denominated facultative because the level of energy introduced by the aerators is sufficient only for oxygenation, but not to maintain the solids (biomass and raw sewage suspended solids) dispersed in the liquid mass. Consequently, the

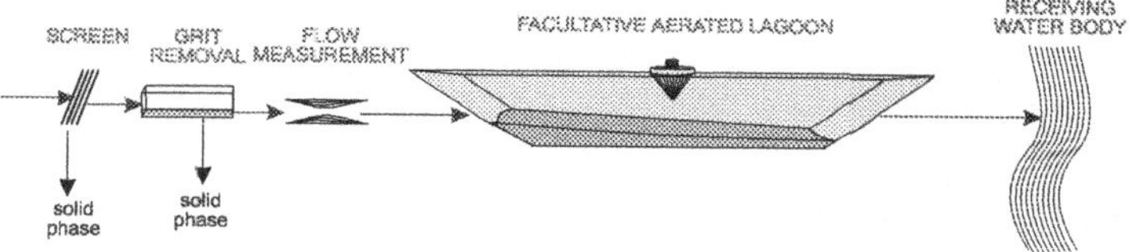

Figure 4.1.   Facultative aerated lagoon system

solids tend to settle and to form a bottom sludge layer, which is decomposed anaerobically. Only the soluble BOD and the BOD represented by finely particulated
solids remain in the liquid mass, undergoing aerobic decomposition. Therefore,
in terms of the distribution of the heterotrophic biomass, the pond behaves as a
conventional facultative pond.

The mechanical aerators more commonly used in the aerated lagoons are high-
speed vertical-shaft units. A greater introduction of oxygen is obtained compared to
the conventional facultative ponds, allowing a faster decomposition of the organic
matter. Consequently, the hydraulic detention time in the pond can be smaller
(of the order of 5 to 10 days), that is, the land requirement is lower.

4.3 DESIGN CRITERIA

The design of facultative aerated lagoons is similar to that of facultative ponds
with respect to the kinetics of BOD removal. There are no requirements in terms
of surface area (surface loading rates), due to the fact that the process is independent
from photosynthesis. Some design criteria are specific for the aeration system, and
are described in Sections 4.5 and 4.6.

The following criteria should be considered:

- *detention time*
- *depth*

a)   Detention time

The detention time should be adopted in order to allow a satisfactory removal
of BOD, according to the kinetics described in Section 4.4a. Usually, the values
adopted vary in the following range:

$$\boxed{t = 5 \text{ to } 10 \text{ d}}$$

b)   Depth

The depth of the pond should be selected in order to satisfy the following criteria:

- compatibility with the aeration system
- need of an aerobic layer of approximately 2 m to oxidise the gases from
  the anaerobic decomposition of the bottom sludge

Usually, the depth varies in the range of:

$$\boxed{H = 2.5 \text{ to } 4.0 \text{ m}}$$

4.4 ESTIMATION OF THE EFFLUENT BOD CONCENTRATION

The estimation of the effluent BOD concentration follows a similar procedure to that used for the facultative ponds (Section 2.6). The influence of the hydraulic regime of the pond should also be taken into consideration, although the formulas corresponding to the complete-mix regime are adopted in most designs. The formulas for the estimation of the dispersion number d, presented in Section 2.6, should not be adopted here, since they are specific for unaerated facultative ponds.

Similarly to the facultative ponds, the effluent from facultative aerated lagoons is constituted of *dissolved organic matter* (**soluble BOD**) and *suspended organic matter* (**particulate BOD**). However, the latter is *not* anymore associated predominantly to algae.

$$\boxed{BOD_{tot} = BOD_{sol} + BOD_{part}} \tag{4.1}$$

where:

BOD_{tot} = total BOD_5 of the effluent (mg/L)

BOD_{sol} = soluble BOD_5 of the effluent (mg/L)

BOD_{part} = particulate (suspended) BOD_5 of the effluent (mg/L)

The *suspended* organic matter is represented mainly by the *bacteria* responsible for the stabilisation of the organic matter. In spite of the fact that facultative aerated lagoons allow the sedimentation of solids, not all of them settle. A large part of the bacterial protoplasm is constituted of organic matter, which exerts an oxygen demand on the receiving body and in the BOD test. In the case of unaerated facultative ponds, the solids in the effluent consist mainly of algae, which may even lead to the production of oxygen in the receiving body. For this reason, the BOD of the effluent from *unaerated facultative ponds* is considered in the European legislation as being mainly the soluble BOD. However, in the case of *aerated lagoons* (and all other wastewater treatment processes, with the exception of unaerated facultative ponds), the BOD related to the organic fraction of the suspended solids (particulate BOD) should be considered.

a) Soluble effluent BOD

The estimation of the soluble effluent BOD is accomplished using the same formulas presented for facultative ponds, which are a function of the hydraulic regime assumed for the reactor (Section 2.6).

The value of the BOD removal coefficient K is higher in the case of facultative aerated lagoons. Typical values for the complete-mix regime are in the range of (Arceivala, 1981):

$$\boxed{K = 0.6 \text{ to } 0.8 \text{ d}^{-1}}$$

This value is for the liquid temperature of 20 °C. For other temperatures, Equation 2.3 can be used, with $\theta = 1.035$.

With relation to the value of S_o to be used in the equations in Table 2.4 and for the design of the aeration system, the following aspects should be taken into consideration. Facultative aerated lagoons allow the sedimentation of the particulate organic matter of the raw sewage, which undergoes *anaerobic* decomposition in the bottom sludge. The influent BOD value (S_o) available for *aerobic* stabilisation is, therefore, lower than the total value in the raw sewage. The value of S_o to be adopted in the calculations depends on the anaerobic activity, which is a function of the temperature in the liquid. Consequently, the following two conditions can happen regarding the organic matter in the bottom sludge (Arceivala, 1981):

* *Anaerobic decay with hydrolysis and acidification, but without methanogenesis*

  $S_o = 100\%$ of total influent BOD

  Climate: cold
  Comment: there are regions with cold periods in which the methanogenic stage (responsible for the removal of BOD) does not fully occurs, implying the release of intermediate by-products of the digestion, which exert an oxygen demand in the aerobic layer. Hence, the BOD to require aerobic stabilisation can be considered as being equal to S_o.
* *Anaerobic decay with hydrolysis, acidification, and methanogenesis*

  $S_o = 40\%$ to 70% of total influent BOD

  Climate: warm
  Comment: under conditions in which the liquid temperature is sufficiently high (>15 °C), the anaerobic conversion is complete, including all the stages. Because a fraction of the BOD is stabilised anaerobically in the bottom, the value of S_o considered for the estimation of the effluent BOD and the oxygen requirements is only a portion of the influent BOD (around 40 to 70%).

However, for *design* purposes, the BOD load to be aerobically stabilised can be considered, for safety reasons, as being equal to the total influent load ($S_o = $ BOD of the influent):

$$\boxed{\text{\textit{Design:} } S_o = 100\% \text{ of total (soluble + particulate) BOD}_5 \text{ of the influent}}$$

b) Particulate effluent BOD

To calculate the particulate effluent BOD from a facultative aerated lagoon it is necessary to estimate the concentration of suspended solids in the effluent from the pond, since the particulate BOD is caused exactly by the suspended solids.

The amount of solids that stay in suspension in the liquid medium is a function of the turbulence level introduced by the aerators. This turbulence level, or mixing capacity, is evaluated through the concept of the **power level**. The power level represents the *energy introduced by the aerators per unit volume of the reactor*, being obtained from the formula:

$$\phi = P/V \tag{4.2}$$

where:

ϕ = power level (W/m^3)
P = power for aeration (W)
V = reactor volume (m^3)

The greater the power level, the greater the quantity of suspended solids that can remain dispersed in the liquid medium (Table 4.1). The values presented in the table are only estimates, since the mixing intensity also depends on the number and distribution of aerators (in the case of mechanical aeration) and on the size and geometry of the pond.

Table 4.1. Suspended solids concentrations that can be maintained dispersed in the liquid as a function of the power level

Power level (W/m^3)	SS (mg/L)
0.75	50
1.75	175
2.75	300

Source: Eckenfelder (1979)

Facultative aerated lagoons work with low power levels, since one of their objectives is exactly to facilitate the sedimentation of the solids. The values of the power level of facultative aerated lagoons are in the range of:

$$\boxed{\text{Power level: } \phi = 0.75 \text{ to } 1.50 \text{ W/m}^3}$$

As a result, the SS concentrations in the lagoon effluent would be in the range of 50 to 140 mg/L. However, the outlet zone of the lagoon may be left without aerators, in order to improve the settling conditions and, therefore, the effluent quality. Usual SS values in the final effluent may then be in the approximate range of:

$$\boxed{\text{SS effluent: 50 to 100 mg/L}}$$

Once the SS concentration in the effluent has been estimated, the calculation of the expected value for the effluent particulate BOD can be undertaken, using the following relationship:

$$\boxed{BOD_{part} = 0.3 \text{ to } 0.4 \text{ mgBOD}_5/\text{mgSS}}$$

Thus, each 1 mg/L of SS produces a particulate BOD between 0.3 and 0.4 mg/L. Knowing the total concentration of effluent SS, the particulate effluent BOD is readily estimated.

4.5 OXYGEN REQUIREMENTS

The amount of oxygen to be supplied by the aerators for the aerobic stabilisation of the organic matter should usually be equal to the total ultimate influent BOD. The ultimate BOD (BOD_u) corresponds to the total oxygen demand exerted for the complete stabilisation of the organic matter. In typical domestic sewage, BOD_u is reached at the end of a long period, in the order of 20 days. BOD_u is therefore higher than the BOD_5, since the latter is exerted only until the fifth day. The ratio BOD_u/BOD_5 is frequently adopted in the range of 1.2 and 1.5.

The considerations made in Section 4.4a are also valid here. Thus, the oxygen demand can be admitted, for *design purposes*, as being due to all the influent BOD, without any reduction related to the anaerobic conversion in the bottom.

In the computation of the oxygen requirements, the following items can be *discounted*:

- Fraction of non-stabilised BOD (S) leaving with the effluent. This is due to the fact that the efficiency of the system in the removal of BOD is lower than 100%. See Section 4.4a. for the estimation of S.
- Fraction of BOD (oxygen consumption) not exerted by the solids leaving with the effluent. This corresponds to the particulate BOD (converted to ultimate BOD), covered in Section 4.4.

Considering these aspects, the amount of oxygen to be supplied can be adopted as:

$$\boxed{OR = a.Q.\,(S_0 - S)/1000} \tag{4.3}$$

where:

    OR = oxygen requirement (kgO_2/d)
     a = coefficient, varying from 0.8 to 1.2 $\text{kgO}_2/\text{kgBOD}_5$
     Q = influent flow (m^3/d)
    S_0 = total (*soluble + particulate*) influent BOD_5 concentration (g/m^3)
     S = *soluble* effluent BOD_5 concentration (g/m^3)
  1000 = conversion from kg to g (g/kg)

4.6 AERATION SYSTEM

The aerators more frequently used for aerated lagoons are the *mechanical vertical-shaft high-speed floating aerators*. Aerators with aspirating devices have also been used.

Both systems require simple maintenance. Installation is also simple, without the need of walkways and supporting columns. If needed, the position of the aerators in the pond can be easily changed. The floating units also adapt themselves to the water level variations in the lagoon, which can be controlled by the outlet weir.

The following aspects should be taken into consideration:

- The aerators should be distributed homogeneously in the aerated zone of the lagoon.
- If the lagoons are predominantly rectangular, a larger number of aerators or more powerful aerators can be placed close to the inlet zone, where the oxygen demand is higher.
- Adjacent aerators should have opposite rotation directions, that is, one should be clockwise and the other anti-clockwise.
- If lower effluent SS concentrations are desired, the final area of the lagoon can be without aerators, in order to provide better settling conditions.
- In small ponds, there should be a minimum of two aerators.
- The manufacturers' data should be consulted with relation to the recommended lagoon depth, influence zone of each aerator, oxygenation efficiency, etc.

There are two types of the area of influence of a mechanical aerator (Figure 4.2):

- *Mixing zone.* Area in which mixing of the liquid is guaranteed, allowing the maintenance of solids in suspension. Area with a smaller diameter, having the aerator in the centre.
- *Oxygenation zone.* Area in which the diffusion of oxygen in the liquid is guaranteed, but not the mixing. Area with a larger diameter, encircling the mixing zone.

Table 4.2 presents approximate values for the operating ranges of mechanical aerators as a function of their power. As can be observed, the area of influence of each aerator for oxygenation is much higher than that for mixing.

4.7 POWER REQUIREMENTS

The required power is calculated based on the oxygen requirements (OR), determined in Section 4.5. The parameter that converts oxygen consumption into power is the *oxygenation efficiency (OE.)*, which is expressed in the units of kgO_2/kWh.

The manufacturers' data are usually expressed in *standard conditions*, to allow a common base for the comparison of the efficiencies. The standard conditions are for *20 °C, absence of dissolved oxygen, no salinity, sea level, clean water*. See Chapter 11 for further information regarding aeration.

Table 4.2. Usual operation ranges of high-speed aerators

Power (HP)	Normal operating depth (m)	Influence diameter (m)		Diameter of the anti erosion plate
		Oxygenation	Mixing	
5–10	2.0–3.6	45–50	14–16	2.6–3.4
15–25	3.0–4.3	60–80	19–24	3.4–4.8
30–50	3.8–5.2	85–100	27–32	4.8–6.0

Notes:

- Usual powers of aerators: 1; 2; 3; 5; 7.5; 10; 15; 20; 25; 30; 40 and 50 HP.
- There are high-speed aerators with greater powers, but they tend to be, overall, less efficient.
- The table presents the influence diameter (and not the radius)
- Anti-erosion plate: situated in the pond bottom, underneath the aerator
- Source: table made based on data presented by Crespo (1995)

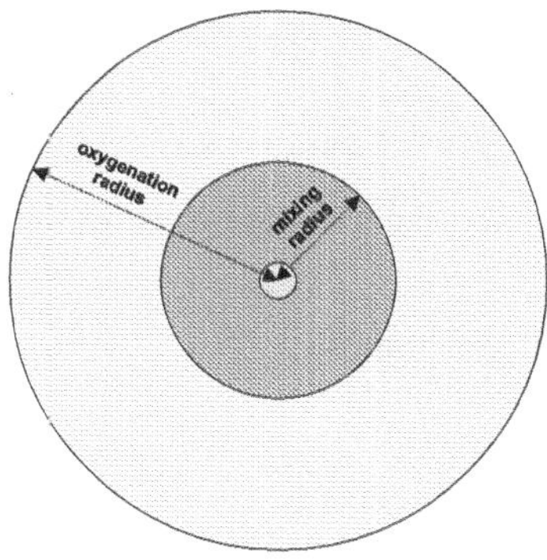

Figure 4.2.   Mixing radius and oxygenation radius in a mechanical aerator

At *standard conditions*, the oxygenation efficiency of the aerators is within the range presented below. However, the manufacturers' data should always be consulted.

$$OE_{standard} = 1.2 \text{ to } 2.0 \text{ kgO}_2/\text{kWh}$$

Under real (field) operating conditions in the treatment plant, the oxygenation efficiency is smaller, being in the following range:

$$OE_{field} = 0.55 \text{ to } 0.65 \; OE_{standard}$$

The power requirements are finally given by the following formula:

$$P = \frac{OR}{24 \times OE_{field}} \tag{4.4}$$

where:

   P = power required (kW)
   24 = conversion from days to hours (24 h/d)

The power of each aerator is then specified based on the manufacturers' data (or Table 4.2). For this, kW needs to be converted into HP (multiply kW by 1.34 to obtain HP).

4.8 SLUDGE ACCUMULATION

The sludge accumulation rate is in the order of 0.03 to 0.08 m^3/inhab.year (Arceivala, 1981). The sludge should be removed when the layer reaches a thickness that can be affected by the aerators, or when the net pond volume is substantially reduced (usually when the sludge reaches 1/3 of the pond depth). The inclusion of grit removal upstream of aerated lagoons is very important.

Example 4.1

Design a facultative aerated lagoon system using the same input data from Example 2.3:

- Population = 20,000 inhabitants
- Influent flow: $Q = 3,000$ m^3/d
- Influent BOD: $S_o = 350$ mg/L
- Temperature: $T = 23$ °C (liquid)

Solution:

a)   Detention time

t = 8 d (adopted)

b)   Effluent soluble BOD

Assuming the complete-mix model and adopting the coefficient $K = 0.7$ d^{-1} for 20 °C, corrected for 0.8 d^{-1} for 23°C:

$$\text{Soluble BOD}_5: S = \frac{S_o}{1 + K.t} = \frac{350}{1 + 0.8 \times 8} = 47 \text{ mg/L}$$

   Lower values of S will be obtained if settling and anaerobic digestion of the influent particulate BOD are considered.

c)   Estimation of the effluent particulate BOD

Assuming that the effluent contains 80 mg/L of suspended solids, the concentration of effluent particulate BOD_5 will be approximately:

Particulate $BOD_5 = 0.35$ mgBOD$_5$/mgSS $\times$ 80 mgSS/L $= 28$ mgBOD$_5$/L

Example 4.1 (Continued)

d)   Total effluent BOD

Total BOD = soluble BOD + particulate BOD = 47 + 28 = 75 mg/L

To reduce the effluent BOD concentration, the detention time could be increased. However, this may not be economical, owing to the need of large volume increases for a small reduction in S. The configuration of the pond could also be changed, approaching a plug-flow reactor. Besides that, the settling conditions in the outlet zone could be improved by the exclusion of some aerators (already done in this example).

The efficiency of the system in the removal of BOD is:

$$E = \frac{S_o - S}{S_o} = \frac{350 - 75}{350} \times 100 = 79\%$$

e)   Required volume

$$V = t.Q = 8\ d \times 3000\ m^3/d = 24,000\ m^3$$

f)   Required area

Adopting a depth H = 3.5 m:

$$A = \frac{V}{H} = \frac{24,000\ m^3}{3.5\ m} = 6,900\ m^2\ (0,69\ ha)$$

g)   Oxygen requirements

$$RO = a.Q.(S_o - S) = \frac{1.0 \times 3000\ m^3/d \times (350 - 47)\ g/m^3}{1000\ g/kg}$$

$$= 909\ kgO_2/d = 38\ kgO_2/h$$

h)   Power requirements

Adopt high-speed floating aerators. The oxygenation efficiency in standard conditions is adopted as:

$$OE_{standard} = 1.8\ kgO_2/kWh$$

The Oxygenation Efficiency in the field can be adopted as around 60% of the standard OE. Thus:

$$OE_{field} = 0.60 \times 1.8\ kgO_2/kWh = 1.1\ kgO_2/kWh$$

Example 4.1 (Continued)

The required power is:

$$P = \frac{OR}{OE} = \frac{38 \text{ kgO}_2/\text{h}}{1.1 \text{ kgO}_2/\text{kWh}} = 34 \text{ kW} \cong 45 \text{ HP}$$

i)   Aerators

Adopt 6 aerators, each of 7.5 HP.
Therefore, the total installed power is 6×7.5 HP $= 45$ HP (34 kW)

j)   Pond dimensions

Adopt two ponds in parallel. With two ponds, there is a larger flexibility during the occasional periods of sludge removal (one pond being cleaned and one pond in operation).

Considering an square area of influence for each aerator, and leaving the final zone without aerators, the pond can have the following dimensions:

Two ponds, each with L $= 116$ m and B $= 29$ m   (8 squares with dimensions
29 m $\times$ 29 m)

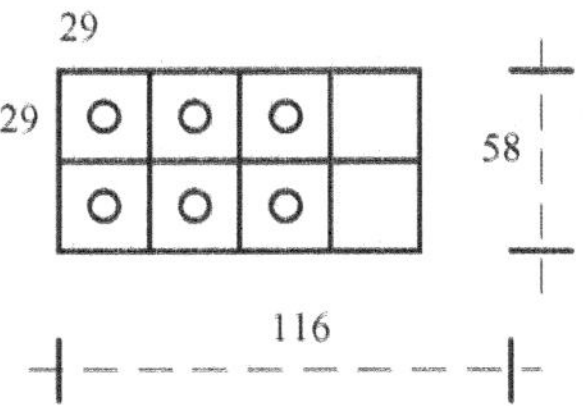

According to Table 4.2, for a power of 7.5 HP for each aerator, the area of influence of 29 m $\times$ 29 m is inside the oxygenation zone (as desired), but is outside the mixing zone (also desirable, for a facultative aerated lagoon).

k)   Verification of the power level

The average power level in the whole lagoon is:

$$\phi = \frac{P}{V} = \frac{34,000 \text{ W}}{24,000 \text{ m}^3} = 1.4 \text{ W/m}^3$$

This power level is expected to maintain solids in suspension. The estimation of 80 mg/l is reasonable (see Table 4.1), considering that there will be some settlement on the unaerated zone of the lagoon. The power level in the aerated zone only is larger, since the volume of the aerated zone is 75% of the total pond volume (3/4 of the pond length have aerators and $^1/_4$ is without aerators – see item j).

Example 4.1 (Continued)

l)   Sludge accumulation

Annual accumulation = 0.05 m^3/inhab.year × 20,000 inhab = 1,000 m^3/year

Thickness in 1 year:

$$\text{Thickness} = \frac{1,000 \text{ m}^3/\text{year} \cdot 1 \text{ year}}{6,900 \text{ m}^2} = 0.14 \text{ m/year}$$

Thickness in 7 years of operation:

Thickness: 0.14 m/year × 7 years = 1.0 m in 7 years

After 7 years of operation, there will be an accumulation of sludge in the order of 1.0 m, which will reduce the net pond depth from 3.5 m to 2.5 m (reduction around 30%). Cleaning will probably be necessary after this period.

m)   Total area required

The required net area is 0.69 ha. The total area required for all the components of the treatment plant is approximately 30% greater than this value. Thus, the total area will be 1.30 × 0.69 = 0.90 ha.

The per capita land requirement is:

$$\text{Per capita land requirement} = \frac{\text{total area}}{\text{population}} = \frac{9,000 \text{ m}^2}{20,000 \text{ inhab}} = 0.45 \text{ m}^2/\text{inhab.}$$

This value is approximately 12% of the value required for a system with facultative ponds only (see Example 2.3).

n)   Arrangement of the system

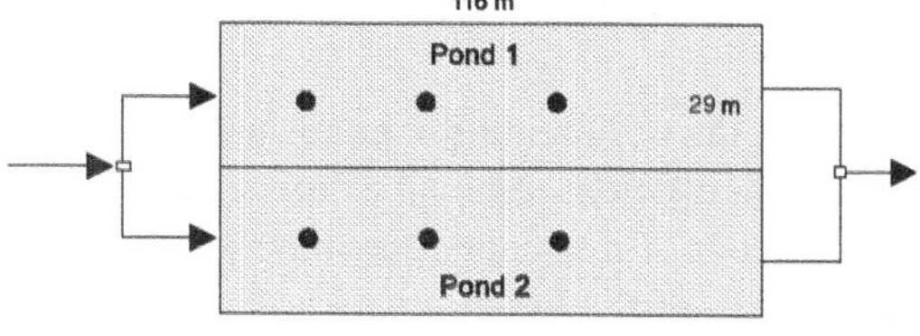

5

Complete-mix aerated lagoons followed by sedimentation ponds

5.1 INTRODUCTION

Complete-mix aerated lagoons are essentially aerobic. The aerators serve not only to *guarantee the oxygenation* of the medium but also to *maintain the suspended solids (biomass) dispersed in the liquid medium.* The typical detention time of a complete-mix aerated lagoon is in the order of 2 to 4 days.

The quality of the effluent from a complete-mix aerated lagoon is not adequate for direct discharge, owing to the high levels of suspended solids. For this reason, these lagoons are usually followed by other ponds, where settling and stabilisation of the settled solids can take place. These ponds are denominated *sedimentation ponds*. Figure 5.1 presents the flowsheet of the system.

The detention times in the sedimentation ponds are low, in the order of 2 days. This time is enough for an efficient removal of the suspended solids produced in the aerated lagoon. However, it does not contribute to an additional biochemical removal of BOD, as a result of the low biomass concentration maintained in suspension in the liquid medium (the biomass tends to settle). Besides this, the sludge accumulation capacity is relatively reduced, implying the need of its removal every 1 to 5 years (there are systems with continuous sludge removal, using pumps coupled to rafts).

The land requirements for this system are the *smallest within the pond systems.* The energy requirements are similar to the other aerated lagoon systems.

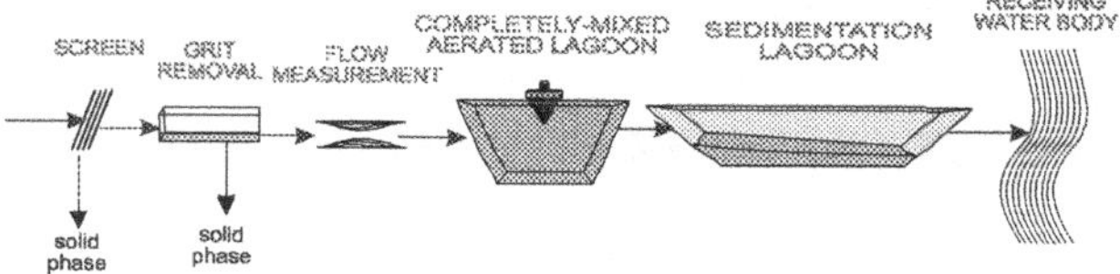

Figure 5.1.   System of complete-mix aerated lagoon followed by sedimentation pond

5.2 DESCRIPTION OF THE PROCESS

In the aerated lagoon, the level of energy introduced by the aerators creates a turbulence that, besides guaranteeing the oxygenation, still allows all the solids to be maintained dispersed in the liquid medium. Therefore, the denomination *complete-mix lagoon* is due to the high degree of energy per unit of volume, responsible for the total mixing of the constituents in the whole pond. These lagoons are also called *flow-through lagoons*, in the sense that the liquid and the solids all flow together in the pond (without solids retention), resulting from the high mixing level. Another designation is *CSTR lagoons* (completely-stirred tank reactor).

Amongst the solids maintained in suspension and in complete mixing are included, besides the organic matter from the raw sewage, also the bacteria (biomass). Consequently, there is a larger concentration of bacteria in the liquid medium, together with a greater organic matter - biomass contact. Thus, the efficiency of the aerobic pond increases, also allowing a reduction in its volume.

The aerated lagoon acts in a similar way to the aeration tanks of the activated sludge process. The main difference is the absence of the recirculation of solids, an essential characteristic of the activated sludge process. Owing to the absence of the recirculation, the concentration of the biomass only reaches a certain value, which is dictated by the availability of the influent substrate (BOD load). The concentration of biological suspended solids in the aerated lagoon is in the order of 20 to 30 times less than in the reactor of activated sludge systems, which justifies the high efficiency of the latter.

However, in spite of the good efficiency of the aerated lagoons in the removal of the organic matter originally present in the wastewater, the quality of their effluent is not satisfactory for direct discharge into the receiving body. The biomass stays in suspension in the whole pond volume, and therefore leaves with the effluent from the aerated lagoon. This biomass is also organic matter, although of a different nature from the BOD of the raw sewage. If this organic matter generated in the lagoon is discharged into the receiving body, it will also exert an oxygen demand, causing the deterioration of the water quality.

Therefore, a downstream unit is needed, in which the suspended solids (predominantly the biomass) can settle. In the present case, this unit is represented by a *sedimentation pond* (in the activated sludge process, it is the secondary sedimentation tank). The effluent from the sedimentation pond leaves with a lower solids level and can be discharged directly into the receiving body.

5.3 DESIGN CRITERIA FOR THE COMPLETE-MIX AERATED LAGOONS

The main design criterion is the *detention time.*

a) Detention time

In complete-mix aerated lagoons, there is the following relationship between the detention times of the liquid and the biomass:

$$\boxed{\text{hydraulic detention time} = \text{solids retention time}}$$

or

$$\boxed{t = \theta_c}$$

The *hydraulic detention time (t) is* the average residence time of the liquid molecules in the reactor. The solids retention time, or *sludge age (θ_c)* is the average residence time of the bacterial cells in the reactor.

In the case of complete-mix aerated lagoons, due to the non-existence of sludge recirculation or any form of solids retention, the molecules of the liquid and the bacterial cells remain the same time in the reactor ($t = \theta c$). This important aspect has hydraulic and process implications. In the activated sludge system, the sludge age is the main design parameter. However, in complete-mix aerated lagoons, the hydraulic detention time (= sludge age) constitutes the main parameter.

In complete-mix aerated lagoons, the detention time varies in the range of:

$$\boxed{t = 2 \text{ to } 4 \text{ d}}$$

If more than one cell in series is adopted, the detention time in each one can be close to 2 days. The advantage of having detention times around 2 days is the reduction in the growth of algae, which could be washed out of the lagoon without being able to develop.

b) Depth

The depth of the pond should be selected in order to satisfy the requirements of the aeration equipment, in terms of mixing and oxygenation.

Usually, depth values are in the range of:

$$\boxed{H = 2.5 \text{ to } 4.0 \text{ m}}$$

5.4 ESTIMATION OF THE EFFLUENT BOD CONCENTRATION FROM THE AERATED LAGOON

For the estimation of the effluent BOD concentration from the aerated lagoon, models similar to those employed for the activated sludge process can be adopted. In this chapter, a simplified version based on first-order reactions is presented. In these conditions, the estimation of the effluent concentration follows a procedure similar to that used for the facultative aerated lagoons (Section 4.4).

The influence of the hydraulic regime of the pond can also be taken into consideration. However, the complete-mix model is usually adopted, since it offers a good approximation to the hydraulic behaviour of this type of aerated lagoon.

Also in this case the effluent from the aerated lagoons is composed of *dissolved organic matter* (**soluble BOD**) and *suspended organic matter* (**particulate BOD**) (see Section 4.4):

$$\boxed{BOD_{tot} = BOD_{sol} + BOD_{part}} \tag{5.1}$$

a)   Soluble effluent BOD

The estimation of the effluent soluble BOD from the aerated lagoon can be done using the same formulas presented for facultative ponds and facultative aerated lagoons, which are a function of the hydraulic regime adopted for the reactor. As commented, the *complete-mix* model can be assumed.

The value of the removal coefficient K is, in the case of complete-mix aerated lagoons, even higher than in the other pond systems. This is due to the larger biomass concentration in the pond. Typical values of K are in the range of (Arceivala, 1981):

$$\boxed{K = 1.0 \text{ to } 1.5 \text{ d}^{-1}}$$

However, this value of K incorporates the influence of the concentration of the volatile suspended solids (VSS or X_v), which represent the biomass. The coefficient K can be dismembered into two fractions, so that:

$$\boxed{K = K'.X_v} \tag{5.2}$$

where:

K' = BOD removal coefficient $(mg/l)^{-1}(d)^{-1}$. The value of K' is in the range of 0.01 to 0.03 $(mg/l)^{-1}(d)^{-1}$ (Arceivala, 1981)

X_v = concentration of volatile suspended solids (mg/L)

According with Equation 5.2, the larger the biomass concentration (X_v), the larger the coefficient K (K' is constant) and, consequently, the larger the BOD removal efficiency.

The effluent soluble BOD concentration from the aerated lagoon is given by:

$$S = \frac{S_o}{1 + K'.X_v.t}$$

(5.3)

As in the other systems, S_o represents the influent total (soluble + particulate) BOD, while S represents the effluent soluble BOD.

It is interesting to point out that, within certain limits, S is independent from the influent concentration S_o. If S_o increases, the biomass concentration (X_v) increases proportionally, due to the larger food availability. If S_o decreases, X_v decreases, and S remains constant. This comment is for steady-state conditions (for design purposes), because fast variations of S_o (typical in operation) are not immediately accompanied by the increase of X_v.

The values of K and K' are for a liquid temperature of $20\,^{\circ}\text{C}$.

The concentration of the biomass (X_v) is a result of the gross growth (positive factor) and the bacterial decay (negative factor). The formula for the calculation of X_v is:

$$X_v = \frac{Y.(S_o - S)}{1 + K_d.t}$$

(5.4)

where:
- Y = yield coefficient ($mgX_v/mgBOD_5$), representing the amount of biomass (mg X_v) that is produced per unit substrate used (mg BOD_5).
- K_d = bacterial decay coefficient or endogenous respiration coefficient (d^{-1}), representing the decay rate of the biomass during endogenous metabolism.

Typical values of these coefficients (Metcalf & Eddy, 1991) are presented in Table 5.1.

The value of the coefficient K_d, in this case, is slightly different from the value of K_d adopted in the chapters relating to activated sludge. In the equations for the activated sludge process a correction is adopted for the biodegradable fraction of VSS, which alter the value of K_d. For simplicity, in the case of aerated lagoons, the formulas are used without the biodegradable fraction concept.

Table 5.1. Kinetic and stoichiometric coefficient values

Coefficient	Unit	Range	Typical value
Y	$mgVSS/mgBOD_5$	0.4–0.8	0.6
K_d	d^{-1}	0.03–0.08	0.06

b)   Particulate effluent BOD

To calculate the effluent particulate BOD from the complete-mix aerated lagoon, it is necessary to estimate the concentration of suspended solids in the effluent from the pond, since this BOD is caused by the suspended solids.

The concentration of volatile suspended solids in the effluent of the aerated lagoon is given by Equation 5.4.

The particulate BOD can be estimated based on the following relationship with the volatile suspended solids:

$$\boxed{\text{BOD}_{part} = 0.4 \text{ to } 0.8 \text{ mgBOD}_5/\text{mgVSS}}$$

In aerated lagoons, the relationship between the volatile suspended solids (VSS or X_v) and the total suspended solids (SS or X) is in the order of:

$$\boxed{X_v/X = 0.7 \text{ to } 0.8}$$

Thus, the particulate BOD can also be estimated as a function of the total suspended solids in the effluent, aggregating the last two relationships:

$$\boxed{\text{BOD}_{part} = 0.3 \text{ to } 0.6 \text{ mgBOD}_5/\text{mgSS}}$$

The particulate BOD in the final effluent is a function of the effluent SS from the sedimentation pond. There are no widely accepted models that allow the estimation of this effluent concentration. For design purposes, an SS removal efficiency around 80 to 85% can be admitted.

5.5 OXYGEN REQUIREMENTS IN THE AERATED LAGOON

The amount of oxygen to be supplied by the aerators for the aerobic stabilisation of the organic matter should usually be equal to the total ultimate BOD (BOD_u) removed (see Section 4.5). The ratio $\text{BOD}_u/\text{BOD}_5$ in the raw wastewater is in the order of 1.2 to 1.5.

In the computation of the total oxygen demand, the consumption not exerted by the volatile suspended solids that leave the system with the effluent can be discounted, similarly to what is done in the activated sludge system calculations. The oxygen requirements can then be calculated by:

$$\boxed{OR = \frac{a.Q.(S_o - S)}{1000}} \tag{5.5}$$

where:
   OR = oxygen requirement (kgO_2/d)
   a = coefficient of oxygen consumption (1.1 to 1.4 $\text{kgO}_2/\text{kgBOD}_5$ removed)

Q = influent flow (m^3/d)

S_o = influent total (*soluble + particulate*) BOD concentration (g/m^3)

S = effluent *soluble* BOD concentration (g/m^3)

1000 = conversion of g to kg (g/kg)

5.6 POWER REQUIREMENTS IN THE AERATED LAGOON

In order to guarantee the mixing energy required for maintaining the suspended solids dispersed in the liquid medium, the *mixing requirements* should be fulfilled. The definition of the power for the aerators is then dictated by the concept of the *power level*.

As seen in Section 4.4b, the power level represents the energy introduced by the aerators per unit reactor volume, and is obtained by:

$$\boxed{\phi = P/V} \tag{5.6}$$

where:

ϕ = power level (W/m^3)

P = power for aeration (W)

V = reactor volume (m^3)

To ensure complete dispersion of the suspended solids in the aerated lagoon, the power level should be:

$$\boxed{\phi \geq 3.0 \ W/m^3}$$

The required power (P) for *mixing* can be calculated through Equation 5.6, by adopting a value for ϕ and knowing V.

The required power for *oxygenation* may be determined using the concepts of Oxygen Requirement (OR) and Oxygenation Efficiency (OE – see Section 4.4).

The installed power must comply with both requirements.

5.7 DESIGN OF THE SEDIMENTATION POND

For the design of the sedimentation pond, the following required volumes should be estimated: (a) volume for clarification (sedimentation) and (b) volume for the storage and digestion of the sludge (Alem Sobrinho and Rodrigues, undated):

Volume required for clarification:

- Detention time: $t \geq 1$ d
- Depth: $H \geq 1.5$ m

Total volume of the pond:

- Detention time (end of the planning horizon): t ≤ 2.0 d (to avoid algal growth)
- Depth: H ≥ 3.0 m (to allow an aerobic layer above the sludge)

The sludge accumulation can be calculated assuming the following data:

- VSS/SS ratio in the influent solids to the settling pond: 0.70 to 0.80 (70 to 80% of the SS are volatile – see Section 5.4b)
- Volatile solids reduction rate in the sludge: $K_v = 0.5$ year^{-1} (50% removal per year) (Arceivala, 1981)

The following equation, modified from Arceivala (1981), allows the estimation of the accumulated sludge volume after a period of t years, as a function of the decay rate of the volatile solids and the accumulation rate of the fixed solids and assuming a density of the sludge close to 1.0:

$$V_t = \frac{\dfrac{M_V}{K_v} \cdot (1 - e^{-K_v \cdot t}) + t \cdot M_F}{1000 \cdot (\text{dry solids fraction})} \qquad (5.7)$$

where:

V_t = volume of sludge accumulated after a period of t years (m^3)

M_V = mass of volatile suspended solids retained in the pond per unit time (kg VSS/year)

M_F = mass of fixed suspended solids retained in the pond per unit time (kg SS$_F$/year)

K_v = decay coefficient of the volatile suspended solids in the sludge in anaerobic conditions (year^{-1}). K_v varies from 0.4 to 0.6 year^{-1}, with an average value of 0.5 year^{-1}

t = time (year)

dry solids = fraction of dry solids in the sludge = 1 – water content fraction in the sludge

---

Example 5.1

Design a complete-mix aerated lagoon followed by a sedimentation pond, using the same data from the previous examples:

- Population = 20,000 inhabitants
- Influent flow: $Q = 3,000$ m^3/d
- Influent BOD: $S_o = 350$ mg/L
- Temperature: T = 23 °C (liquid)

Example 5.1 (*Continued*)

Solution:

Aerated lagoon

a)   Adoption of the detention time

t = 3 d (adopted)

b)   Required volume

$V = t.Q = 3\ d \times 3,000\ m^3/d = 9,000\ m^3$

c)   Required area

Adopting a depth H = 3.5 m:

$$A = \frac{V}{H} = \frac{9,000\ m^3}{3.5\ m} = 2{,}570\ m^2$$

The dimensions of the pond can be:

$$50\ m \times 50\ m(0.25\ ha)$$

d)   Estimation of the concentration of volatile suspended solids (VSS) in the aerated lagoon

Kinetic coefficients (see Table 5.1):

- $Y = 0.6$ (adopted)
- $K_d = 0.06$ (adopted)

Estimation of the effluent soluble BOD concentration (S):

$$S = 50\ mg/L\ \text{(initial estimate)}$$

$$X_v = \frac{Y.(S_o - S)}{1 + K_d.t} = \frac{0.6 \times (350 - 50)}{1 + 0.06 \times 3} = 153\ mg/L$$

e)   Estimation of the effluent soluble BOD

Assuming the complete-mix regime, and adopting the coefficient $K' = 0.017\ (mg/L)^{-1}(d)^{-1}$, which corresponds to $0.015\ (mg/L)^{-1}(d)^{-1}$ for 20 °C, after correction for 23 °C:

$$\text{Soluble BOD}_5: S = \frac{S_o}{1 + K'.X_v.t} = \frac{350}{1 + 0.017 \times 153 \times 3} = 40\ mg/L$$

Example 5.1 (Continued)

In item d above, the initial estimate of S = 50 mg/L can be corrected to S = 40 mg/L, and the VSS concentration recalculated until a satisfactory convergence. However, in this case, the differences will be small (for S = 40 mg/L $\rightarrow X_v$ = 158 mg/L).

This value of the soluble BOD is for the effluent from the aerated lagoon, as well as for the final effluent (since the removal of soluble BOD is neglected in the sedimentation pond).

f)   Estimation of the effluent particulate BOD

Considering that the effluent from the aerated lagoon contains 153 mg/L of volatile suspended solids, the effluent particulate BOD from the aerated lagoon will be:

$$BOD_5 \text{ part} = 0.6 \text{ mgBOD}_5/\text{mgVSS} \times 153 \text{ mgVSS/L} = 92 \text{ mgBOD}_5/L$$

This value is high for direct release into the receiving body, which justifies the need of the sedimentation pond downstream. Assuming that the sedimentation pond presents an efficiency of 85% in the removal of these volatile suspended solids, the VSS concentration in the final effluent from the system will be:

$$VSS_e = \frac{(100 - E)}{100}.VSS_o = \frac{(100 - 85)}{100}.153 = 23 \text{ mg/L}$$

Thus, the particulate BOD in the final effluent will be:

$$BOD_5 \text{ part} = 0.6 \text{ mgBOD}_5/\text{mgVSS} \times 23 \text{ mgVSS/l} = 14 \text{ mgBOD}_5/L$$

g)   Effluent total BOD

Total BOD = soluble BOD + particulate BOD = 40 + 14 = 54 mg/L

The efficiency of the system in the removal of BOD is:

$$E = \frac{S_o - S}{S_o} = \frac{350 - 54}{350}.100 = 85\%$$

h)   Oxygen requirements

The oxygen requirements are around 1.1 to 1.4 of the removed BOD_5 load. Adopting the value of 1.2 $\text{kgO}_2/\text{kgBOD}_{rem}$:

$$RO = a.Q.(S_o - S) = \frac{1.2 \times 3000 \text{ m}^3/\text{d. } (350 - 40) \text{ g/m}^3}{1000 \text{ g/kg}} = 1116 \text{ kgO}_2/\text{d}$$

$$= 47 \text{ kgO}_2/\text{h}$$

Example 5.1 (Continued)

i)   Energy requirements

Adopt high-speed floating mechanical aerators. The Oxygenation Efficiency OE, in standard conditions, is in the order of:

$$OE = 1.8 \text{ kgO}_2/\text{kWh}$$

The oxygenation efficiency in the field can be adopted as around 60% of the standard OE. Thus:

$$OE_{field} = 0.60 \times 1.8 \text{ kgO}_2/\text{kWh} = 1.1 \text{ kgO}_2/\text{kWh}$$

The required power is:

$$P = \frac{OR}{OE} = \frac{47 \text{ kgO}_2/\text{h}}{1.1 \text{ kgO}_2/\text{kWh}} = 43 \text{ kW} = 57 \text{ HP}$$

j)   Aerators

Adopt four aerators, each of 15 HP.
Therefore the total installed power is 4×15 HP $= 60$ HP (45 kW)
 Each aerator will be responsible for an area of influence of 25 m $\times$ 25 m (the dimensions of the pond are 50 m $\times$ 50 m).
 According with Table 4.2, for the power of 15 HP, the influence area is inside the oxygenation zone and close to the mixing zone. The depth of the pond is also satisfactory.

k)   Verification of the power level

$$\phi = \frac{P}{V} = \frac{45,000 \text{ W}}{9,000 \text{ m}^3} = 5.0 \text{ W/m}^3$$

 This power level is enough to maintain all the solids in suspension (Table 4.1). Besides, it is greater than the value of 3.0 W/m^3 suggested as the minimum for complete-mix aerated lagoons.

Sedimentation pond

l)   Design of the sedimentation pond

- Clarification zone (reserved for the liquid):
  Detention time: t $= 1.0$ d (adopted)
  Volume: $V_{clarif} = t.Q = 1.0$ d $\times 3,000$ m^3/d $= 3,000$ m^3
  Depth: $H_{clarif} = 1.5$ m (adopted)
  Required area:

$$A = \frac{V}{H} = \frac{3,000 \text{ m}^3}{1.5 \text{ m}} = 2,000 \text{ m}^2 (0.20 \text{ ha})$$

Example 5.1 (Continued)

- Sludge zone (reserved for the storage and digestion of the sludge):
  Add an additional depth of 1.5 m.
- Total dimensions and values (clarification and sludge zones):
  Total area: 2,000 m^2
  Depth: 1.5 m + 1.5 m = 3.0 m
  Total volume: 2000 m^2 × 3.0 m = 6,000 m^3
  Number of ponds: 2
  Dimensions of each pond: 40 m × 25 m × 3.0 m
  Detention time in a still clean pond:

$$t = \frac{V}{Q} = \frac{6,000}{3,000} = 2.0\,d$$

m)   Sludge accumulation

The load of influent solids to the settling pond is composed of volatile suspended solids VSS (determined in Section d) and fixed suspended solids SS_F. Assume a ratio of 0.75 for VSS/SS (see Section 5.4.b). Hence, the ratio SS_F/VSS will be:

$$SS_F/VSS = (1 - 0.75)/0.75 = 1/3$$

The influent solids loads to the pond per year are:

Volatile solids: VSS = 3,000 m^3/d × 0.153 kgVSS/m^3 × 365 d/year = 167,535 kgVSS/year
Fixed solids: SS_F: =3,000 m^3/d × (0.153/3) kgSS$_F$/m^3 × 365 d/year = 55,845 kgSS$_F$/year

Assuming a removal of 85% of the solids in the settling pond, the loads of volatile and fixed suspended solids that will be added to the sludge layer in the pond are:

M_v = 0.85 × 167,535 = 142,405 kgVSS/year
M_F = 0.85 × 55,845 = 47,468 kgSS$_F$/year

Adopting Equation 5.7 for the estimation of the sludge accumulation after a period of t years, and assuming a fraction of dry solids in the sludge of 8% (water content = 92%):

$$V_t = \frac{\frac{M_v}{K_v}.(1 - e^{-K_v.t}) + t.M_F}{1000.(\text{dry solids content})} = \frac{\frac{142,405}{0.5}.(1 - e^{-0.5 \times t}) + t \times 47,468}{1000 \times 0.08}$$

Example 5.1 (Continued)

For different values of t, the sludge accumulation is:

Time (years)	Accumulated volume (m³)	Ratio V_{sludge}/V_{pond} $= H_{sludge}/H_{pond}$	Sludge height (m)
0.5	1082	0.18	0.54
1.0	1991	0.33	0.99
1.5	2765	0.46	1.38
2.0	3433	0.57	1.71
2.5	4020	0.67	2.01
3.0	4542	0.76	2.28
3.5	5015	0.84	2.52

Column 2: equation above

Column 3: (column 2)/6,000 m³, where 6,000 m³ is the volume of the sedimentation pond

Column 4: (column 3) × 3.0 m, where 3.0 m is the total height of the sedimentation pond

It is observed that after a period of around 1.7 years of operation, the volume reserved for sludge accumulation (corresponding to the height of 1.5 m) is totally used. Therefore, the removal of the sludge from the pond is necessary before this period.

After 1.5 years, the volume of accumulated sludge corresponds to the following accumulation rate per inhabitant per year:

(2,765 m³/ 1.5 years) / 20,000 inhab. = 0.09 m³/inhab.year

n)   Total area required (aerated lagoon + sedimentation pond)

Total area = 0.25 + 0.20 = 0.45 ha

The total area required for all the components of the works is approximately 30% higher than this value. Thus, the total area will be 1.30 × 0.45 ha = 0.59 ha (<0.90 ha, area required for the facultative aerated lagoon - Example 4.1).

The per capita area requirement is:

$$\text{Per capita land requirement} = \frac{\text{total area}}{\text{population}} = \frac{5,900 \, \text{m}^2}{20,000 \, \text{inhab.}} = 0.30 \, \text{m}^2/\text{inhab.}$$

This requirement is twelve times less than that for a primary facultative pond (Example 2.3).

Example 5.1 (Continued)

o)   Arrangement of the system

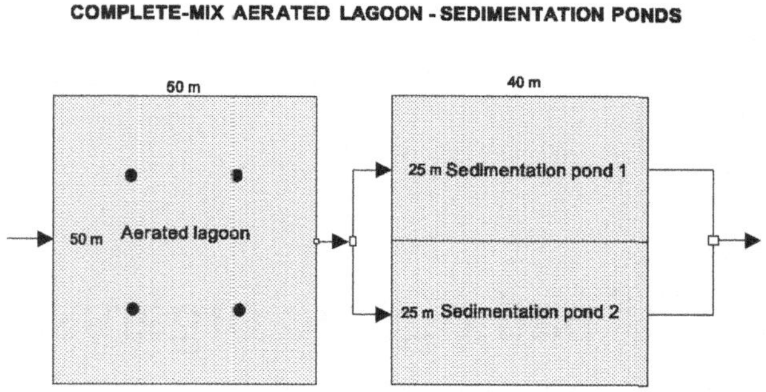

6

Removal of pathogenic organisms

6.1 INTRODUCTION

The removal of pathogenic organisms is one of the most important objectives of stabilisation ponds. The organisms to be removed include bacteria, viruses, protozoan cysts and helminth eggs. A certain removal occurs in the anaerobic, facultative and aerated ponds. However, most of the removal takes place in the **maturation ponds**, which are especially designed for this purpose. Table 1.3 in Chapter 1 presents a summary of the removal efficiencies of the pathogens of interest in the main stabilisation pond systems.

Maturation ponds lead to a polishing of the effluent from any of the stabilisation pond systems previously described or, in broader terms, from any wastewater treatment system. Figure 6.1 shows the flowsheet of a system of anaerobic-facultative ponds followed by a series of maturation ponds. The main objective of maturation ponds is the **removal of pathogens**, and not an additional BOD removal. Maturation ponds constitute an economic alternative to the disinfection of the effluent by more conventional methods, such as chlorination.

6.2 PROCESS DESCRIPTION

The ideal environment for pathogenic organisms is the human intestinal tract. Outside it, in the sewerage system, sewage treatment plant or in the receiving water body, the pathogenic organisms tend to die. Several factors contribute to the

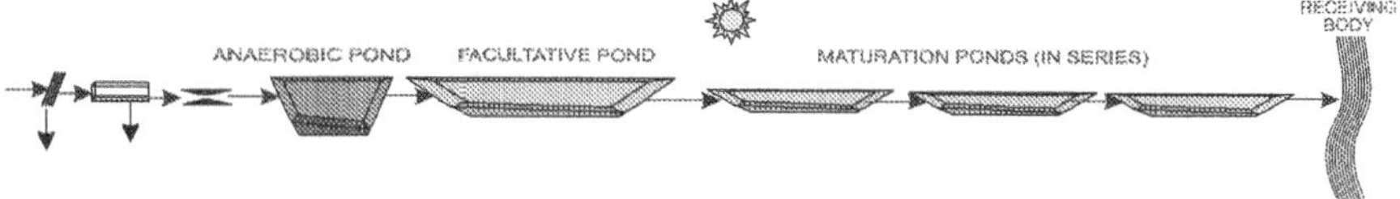

Figure 6.1. Typical flowsheet of a system of stabilisation ponds followed by maturation ponds in series.

removal of the pathogenic organisms:

- *bacteria* and *viruses*: temperature, solar radiation, pH, food shortage, predator organisms, toxic compounds
- *protozoan cysts* and *helminth eggs*: sedimentation

Maturation ponds are designed in order to provide an optimal utilisation of these mechanisms, especially for the removal of *bacteria* and *viruses*, which can be represented by the *coliforms* as indicators. Some of these mechanisms are more effective with smaller pond depths, which justifies the fact that the maturation ponds are shallower, compared with other types of ponds. Among the mechanisms associated to the low *depth* of the pond, the following can be mentioned (van Haandel et Lettinga, 1994; van Buuren et al, 1995; Cavalcanti et al, 2001):

- *High penetration of the solar radiation* (ultraviolet radiation)
- *High pH* (due to high photosynthetic activity)
- *High DO concentration* (favouring the aerobic community, which is more efficient in the removal of coliforms, besides increasing the removal rate due to other mechanisms, such as photooxidation)

The *maturation ponds* should reach high *coliform* removal efficiencies (E > **99.9** or **99.99**%), so that the effluent can comply with most uses of the water in the receiving water body, or for direct uses, such as irrigation (see Section 6.4). In order to maximise the coliform removal efficiency, the maturation ponds are designed with one of the following two configurations: (a) three or four ponds in series or (b) a single pond with baffles. These aspects will be detailed in this chapter.

Regarding the other organisms of public health importance, which are not well represented by coliforms as indicators, the ponds usually reach complete (**100%**) removal of *protozoan cysts* and *helminth eggs* (Arceivala, 1981). The major removal mechanism is sedimentation.

6.3 ESTIMATION OF THE EFFLUENT COLIFORM CONCENTRATION

6.3.1 Influence of the hydraulic regime

The decay of the pathogenic organisms (bacteria and viruses), as well as of the indicators of faecal contamination (coliforms), follows *first-order kinetics* (similarly to

Table 6.1. Formulas for the calculation of the effluent coliform concentration (N)
from ponds

Hydraulic regime	Scheme	Formula for the effluent coliform concentration (N)
Plug flow		$N = N_o e^{-K_b.t}$
Complete mix (1 cell)		$N = \dfrac{N_o}{1 + K_b.t}$
Complete mix (equal cells in series)		$N = \dfrac{N_o}{(1 + K_b.t/n)^n}$
Dispersed flow		$N = N_o \cdot \dfrac{4ae^{1/2d}}{(1 + a)^2 e^{a/2d} - (1 - a)^2 e^{-a/2d}}$ $a = \sqrt{1 + 4K_b.t.d}$

N_o = coliform concentration in the influent(org/100mL) $\quad$ t = detention time (d)
N = coliform concentration in the effluent(org/100mL) $\quad$ n = number of ponds in series (−)
K_b = bacterial die − off coefficient (d^{-1}) $\quad$ d = dispersion number (dimensionless)

the BOD stabilisation in the pond systems, which also follows first-order kinetics).
According with the first-order reactions, the *die-off rate of pathogens is proportional to the pathogen concentration at any time.* Hence, the greater the pathogen concentration, the larger the die-off rate. A similar comment is valid for the coliforms.

Therefore, the same considerations made in Section 2.6 are valid here. The hydraulic regime of the ponds has a great influence in the coliform removal efficiency. The decreasing order of efficiency is:

− *plug-flow pond*	greater efficiency
− *complete-mix ponds in series*	⇓
− *single complete-mix pond*	lower efficiency

Table 6.1 presents the formulas used for the determination of the coliform count in the effluent from ponds, as a function of the different hydraulic regimes.

6.3.2 Idealised hydraulic regimes

In order to obtain the extremely high coliform removal efficiencies that are usually required, the adoption of cells in series or a reactor approaching plug flow (theoretically equivalent to an infinite number of cells) is necessary. Table 6.2 presents the theoretical relative reactor volumes required, as a function of the number of cells, so that the same efficiency is reached. All the values are expressed as a function of the dimensionless product $K_b.t$. Thus, for a certain value of K_b, different total detention times are given, or, in other words, the total relative volume required. If the value of K_b is known, the table can be used for the direct calculation of the

Table 6.2. Theoretical relative volumes necessary to reach a certain removal efficiency, as a function of the number of complete-mix ponds in series

Number of ponds in series	Relative volume (dimensionless product $K_b.t$)			
	E = 90%	E = 99%	E = 99.9%	E = 99.99%
1	9.0	99	999	9999
2	4.3	18	61	198
3	3.5	11	27	62
4	3.1	8.6	18	36
5	2.9	7.6	15	27
∞ (plug flow)	2.3	4.6	6.9	9.2

total volume required (calculation of t, followed by the calculation of V, knowing that $V = t.Q$).

The interpretation of Table 6.2 leads to the following comments:

- with only one ideal complete-mix pond, extremely high volumes are necessary to reach satisfactory coliforms removal (for E = 99.99%, the necessary volume is approximately 1.000 times greater than for an ideal plug-flow reactor)
- with ponds in series, a substantial reduction of volume occurs only with a system comprised of more than 3 cells
- the ideal plug-flow reactor requires small volumes in comparison to the other systems
- these comments are valid assuming the ponds to be ideal reactors (what does not strictly occurs, in practice – plug-flow conditions are seldom achieved in practice)

Figure 6.2 illustrates the efficiencies and the number of logarithmic units removed, for different values of the dimensionless pair $K_b.t$ and the number of ideal complete-mix cells in series. An efficiency of E = 90% corresponds to the removal of one logarithmic unit; E = 99% → 2 log units; E = 99.9% → 3 log units; E = 99.99% → 4 log units; E = 99.999% → 5 log units, and so on, according to the formula:

$$\boxed{\text{log units removed} = -\log_{10}[(100 - E)/100]} \qquad (6.1)$$

In the figure, the highest efficiency of the ideal plug-flow reactor is again seen. Removal efficiencies above 99.9% without excessively large detention times can only be reached with a number of cells in series greater than four or preferably with a plug-flow regime.

However, it should be commented that *plug flow is an idealised hydraulic regime. In practice, it can be only approached (but not reached) through the adoption of*

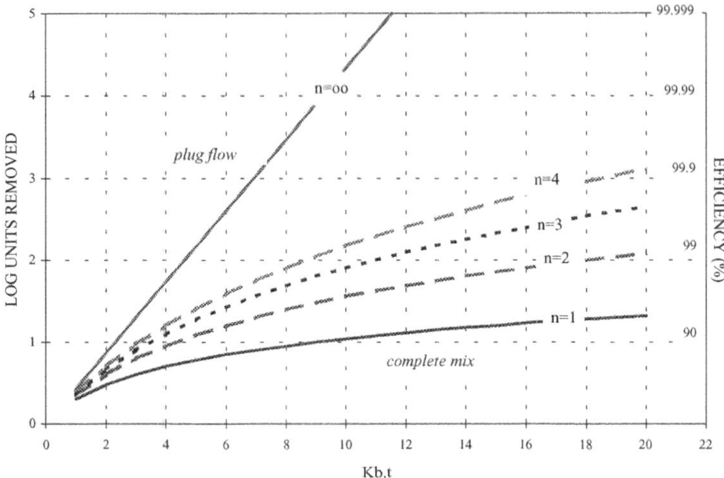

Figure 6.2.   Coliform removal efficiencies, for different values of K_b.t and number of cells in series, assuming the complete-mix hydraulic regime

a low dispersion, induced by baffles. Zero dispersion (as assumed in the plug flow regime) is hardly achievable in a pond.

6.3.3 The dispersed-flow hydraulic regime

In reality, the behaviour of ponds follows the dispersed-flow hydraulic regime, and not the idealised regimes of complete mix and plug flow. Figure 6.3 presents the graph of the values of the efficiency E and the number of logarithmic units removed as a function of the dimensionless pair K_b.t and the dispersion number d. The determination of the dispersion number d was discussed in Section 2.6. It should be borne in mind that the coefficient K_b in the dispersed-flow regime is usually different from the value adopted for the complete-mix regime (see Sections 6.3.4 and 2.6.4).

In the case of a single pond, the figure shows clearly the importance of having a pond with a low dispersion number, tending to the plug-flow regime, in order to increase the removal efficiency. To obtain efficiencies greater than 99.9% (3-log removal) without excessive detention times, a dispersion number lower than 0.3, or preferably 0.1, is needed. These dispersion numbers are only obtained in ponds that have a length/breadth (L/B) ratio greater than 5 or 10 (see Table 2.7).

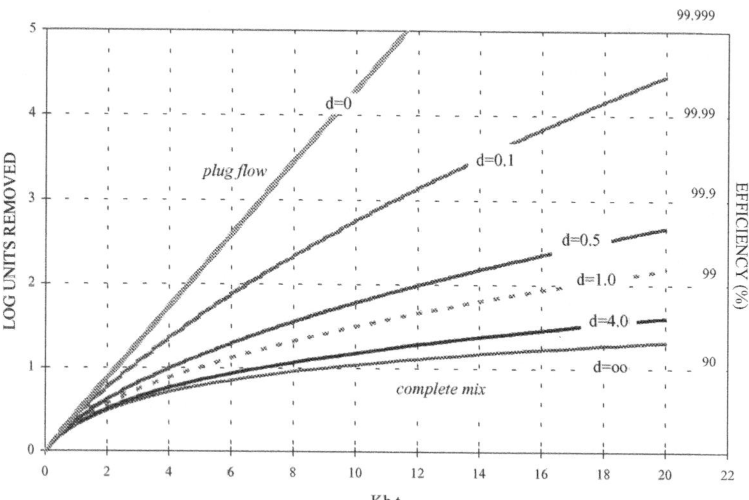

Figure 6.3.   Coliform removal efficiency and number of log units removed in a single pond, for different values of $K_b.t$ and d, assuming the dispersed-flow hydraulic regime

Figure 6.4 presents the number of logarithmic units removed and the removal efficiency in maturation ponds, expressed as a function of the length / breadth (L/B) ratio. In this figure, the relationship between the L/B ratio and the dispersion number d was calculated using the equation d = 1/ (L/B) (Equation 2.14).

The calculation of the L/B ratio in a pond with internal divisions (baffles) can be approximated by:

- *divisions parallel to the breadth B:*

$$L/B = \frac{B}{L}(n+1)^2 \qquad\qquad (6.2)$$

- *divisions parallel to the length L:*

$$L/B = \frac{L}{B}(n+1)^2 \qquad\qquad (6.3)$$

where:

    L/B = resultant internal length/breadth ratio in the pond
    L = length of the pond (m)
    B = breadth of the pond (m)
    n = number of internal divisions

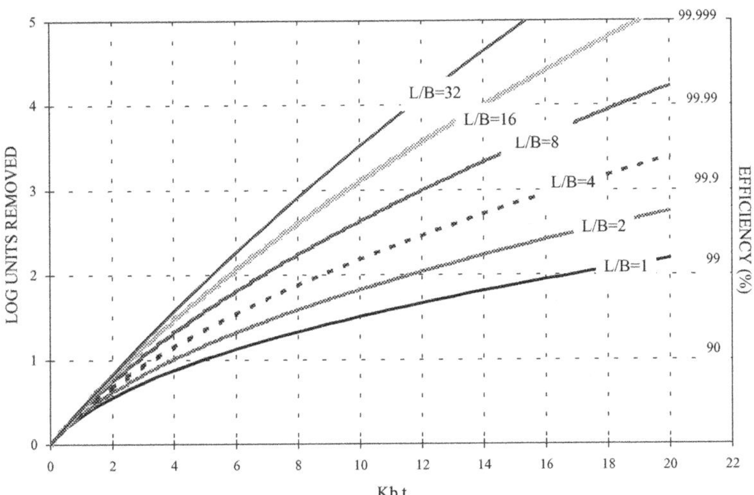

Figure 6.4.   Coliform removal efficiency and number of log units removed for different values of K_b.t and L/B ratio, assuming dispersed flow. The relationship between L/B and d was calculated according to d = 1/ (L/B) (Equation 2.14).

6.3.4 The coliform die-off coefficient K_b according to the dispersed-flow regime

The coliform die-off coefficient (K_b) has a great influence on the estimation of the effluent coliform concentration. The literature presents a great scatter of reported coefficients, together with the additional complication that the different values of K_b have been obtained assuming different hydraulic regimes (not always reported). Besides that, there are other influencing factors, such as DO concentration, pH, solar radiation, BOD loads and the physical configuration of the pond.

The depth exerts a great influence in K_b: shallower ponds have higher K_b values because of the following points: (a) higher photosynthetic activity throughout the pond depth, leading to high pH and DO values; (b) higher penetration of the UV radiation throughout the pond depth (Catunda et al, 1994; van Haandel and Lettinga, 1994; von Sperling, 1999). However, the combined effect of the shallower ponds should be analysed: K_b is larger, but the detention time t is smaller (for a given surface area). The impact on the product K_b.t can be evaluated through the formulas presented for the different hydraulic regimes.

In ponds located in warm-climate regions and with a tendency to stratification, the anaerobic layer at the bottom plays a negative role. The bacterial die-off in anaerobic conditions is lower than in aerobic conditions. Therefore, in a facultative pond, the coliform removal efficiency in the summer may be lower than in a mild winter, in which there is a larger predominance of the aerobic conditions (Arceivala, 1981).

In a review of the international literature, von Sperling (1999) identified values of K_b for facultative and maturation ponds varying from 0.2 to 43.6 d^{-1} (20 °C). This is an extremely wide range, which gives little reliability for design purposes. The highest values were due to the fact that, in case the complete-mix regime had been assumed for a pond that did not behave in practice as an ideal complete mix, there was a tendency of obtaining overestimated values of K_b.

Von Sperling (1999) investigated data from **33** *facultative* and *maturation* ponds in Brazil. The ponds analysed were distributed from the Northeast (latitude 7 ° S) to the South (latitude 23.5 ° S) of the country, covering a tropical to subtropical range of climates. The ponds had different volumes and physical configurations, with 13 being pilot units and the other 20 in full scale. The ponds represented a wide spectrum of operational conditions, with the length / breadth ratio (L/B) varying from 1 to 142 and the detention times from 0.5 to 114 days. In most cases, the coliform removal efficiency was based on average or long-term geometric means. The total number of data used was 66.

Complete-mix and dispersed-flow regimes were analysed in the work. It was observed that the values of the coefficient K_b for dispersed flow were related to the depth of the pond and to the hydraulic detention time. The lower the depth and the detention time, the larger the value of the coefficient K_b. As mentioned, the influence of the smaller depths is a result of the larger penetration of sunlight in the whole water mass (larger photosynthesis, larger dissolved oxygen, and larger pH values), besides the greater penetration of the ultraviolet radiation, which is bactericide. No significant relationship was observed between K_b and the depth or detention time for the complete-mix model.

An equation correlating K_b (dispersed flow) with the depth and the hydraulic detention time was determined through non-linear regression analysis with the available data (von Sperling, 1999):

$$\boxed{K_b \text{ (dispersed)} = 0.917.H^{-0.877}.t^{-0.329} \qquad \textit{(33 ponds in Brazil)}} \qquad (6.4)$$

The Coefficient of Determination was very high ($R^2 = 0.847$), indicating a good fitting of the proposed model to the experimental data. Even though it was known, *a priori*, that a model with such a simple structure would have difficulty in reproducing the wide diversity of situations that occur in practice, there was the advantage of depending only on variables which, in a design application, are known beforehand (H and t). Some of the models available in the literature are

Table 6.3. Values of K_b (dispersed flow), obtained from Equation 6.5 ($K_b = 0.542.H^{-1.259}$), for facultative and maturation ponds

H (m)	0.6	0.8	1.0	1.2	1.4	1.6	1.8	2.0	2.2	2.4
K_b (d^{-1})	1.03	0.72	0.54	0.43	0.35	0.30	0.26	0.23	0.20	0.18

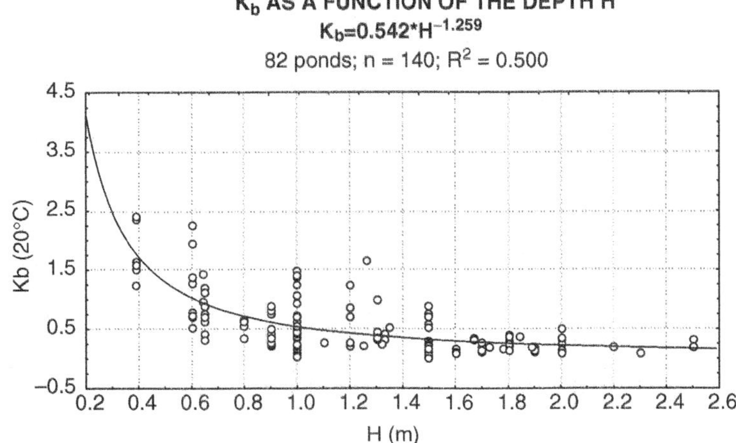

Figure 6.5. Regression analysis between K_b (20 °C, dispersed flow) and the depth H of the ponds. Dispersion number adopted as d = 1/(L/B). 140 results from 82 facultative and maturation ponds in the world.

less practical, because they depend on variables that are not known at the design stage. In spite of the limitations, the model lead to a very good prediction of the logarithm of the effluent coliform concentrations from the 33 ponds ($R^2 = 0.959$).

Subsequently, the author enlarged the database to 82 ponds (140 mean data) in Brazil and in other countries (Argentina, Colombia, Chile, Venezuela, Mexico, Spain, Belgium, Morocco and Palestine). Equation 6.4 was still shown to be valid, although the Coefficient of Determination was reduced to $R^2 = 0.505$. In this enlarged data set, it was observed that the hydraulic detention time exerted a smaller influence and that it could be removed from the equation, without significantly affecting the performance of the model. The new equation obtained is presented below (see also Figure 6.5 and Table 6.3, showing the values of K_b and the best-fit curve). The prediction of the log of the effluent coliform concentration was still entirely satisfactory.

$$K_b \text{ (dispersed)} = 0.542.H^{-1.259} \qquad \textit{(82 ponds in the world)} \qquad (6.5)$$

To allow a better visualisation of the results from both equations (Equations 6.4 and 6.5), Figure 6.6 presents the resulting curves for detention times varying

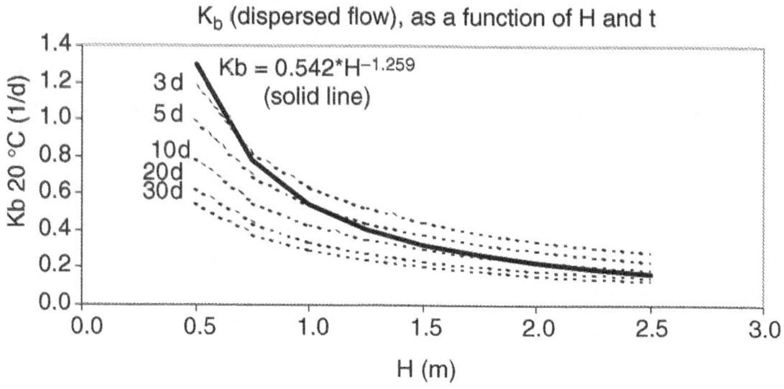

Figure 6.6.   Relation between K_b, H, and t, according to the models proposed for K_b (20 °C, dispersed flow), for facultative and maturation ponds. Dashed curves: Equation 6.4 (33 ponds in Brazil); solid curve: Equation 6.5 (82 ponds in the world).

from 3 to 30 days, and depths varying from 0.5 to 2.5 m. It can be observed that the simpler model (Equation 6.5), based only on the depth H of the pond, is situated in an intermediate range between the curves of the model based on H and t (Equation 6.4), especially for depths greater than 1.0 m. For depths lower than 1.0 m, Equation 6.5 approaches Equation 6.4 only for low values of the hydraulic detention time. Low values of H and t occur simultaneously in maturation ponds in series, which also justifies that the simpler model keeps its practical applicability also for this range of values of H and t.

With the 140 data from the 82 facultative and maturation ponds in the world, it was tested whether the position of the pond in the series would have any influence on the coefficient K_b. The reason is due to the fact that primary and possibly secondary ponds tend to receive a higher BOD surface loading rate, not being, therefore, optimised for the production of high DO and pH values, as in tertiary and subsequent ponds. Even though an statistically significant difference has not been detected, if a refinement in the calculation is desired, the data suggest the following corrections in the values obtained from Equation 6.5 ($K_b = 0.542.H^{-1.259}$):

- *Primary and secondary ponds* $-K_b$: 5 to 15% lower than the value from the general equation
- *Tertiary and subsequent ponds* $-K_b$: 5 to 15% higher than the value from the general equation

Although Equation 6.5 has been derived from a large number of ponds distributed in several places of the world, specific local conditions can always prevail and lead to different values of K_b. For instance, places with very high solar radiation are prone to having high K_b values (higher UV radiation, higher photosynthesis, higher DO and higher pH). As mentioned, to incorporate this and other factors in the equation would lead to a very sophisticated model structure, requiring input data difficult to obtain in practice.

6.3.5 The coliform die-off coefficient K_b according to the complete-mix regime

In spite of the great advantages widely recognised for the dispersed-flow model, it is accepted that the idealised complete-mix model has been more utilised by designers. Von Sperling (2002) analysed the theoretical relationship between the coefficients, according to the hydraulic regimes of complete mix and dispersed flow, and proposed equations, based on regression analysis, which lead to an easy conversion between them. The equations allow the estimation of K_b for the complete-mix regime, based on the coefficient K_b for dispersed flow, on the detention time t (product $K_{b\,disp}.t$) and the dispersion number d. Two equations have been proposed, with different applicability ranges: one for a narrower range (more accurate in this narrow range) and another for a wider range of $K_b.t$ and d, covering most of the ponds found in practice:

Wider range (d varying from 0.1 to 4.0; $K_{disp}.t$ varying from 0 to 10):

$$\frac{K_{mix}}{K_{disp}} = 1.0 + \left[0.0020 \times (K_{disp}.t)^{3.0137} \times d^{-1.4145}\right] \qquad (6.6)$$

Narrower range (d varying from 0.1 to 1.0; $K_{disp}.t$ varying from 0 to 5):

$$\frac{K_{mix}}{K_{disp}} = 1.0 + \left[0.0540 \times (K_{disp}.t)^{1.8166} \times d^{-0.8426}\right] \qquad (6.7)$$

where:

K_{disp} = bacterial die-off coefficient according to the dispersed flow regime (d^{-1})
K_{mix} = bacterial die-off coefficient according to the complete-mix regime (d^{-1})

These equations are valid, not only for coliforms, but also for other constituents that follow first-order kinetics, such as BOD.

The coefficient K_b for complete mix can be obtained from Equations 6.6 or 6.7, within the applicability range of each equation. It may be observed in both equations that, due to the factor of 1.0 on the right-hand side, *the coefficient for complete mix will always be greater than that for dispersed flow.*

The coefficient K_b for dispersed flow can be obtained from Equations 6.4 or 6.5. The dispersion number can be obtained from the formulas presented in Chapter 2 (Polprasert & Batharai, 1983; Agunwamba et al, 1992; Yanez, 1993; von Sperling, 1999). However, it is believed that the formula d = 1/(L/B) (von Sperling, 1999) (Equation 2.14) can be adopted, given its simplicity and similarity of results with the other formulas.

It should be highlighted that, in principle, the die-off coefficient should not vary with the hydraulic model, but only represent the coliform decay according to its kinetics (as determined in a batch test). However, the inadequacy of the idealised hydraulic regimes in representing in a perfect way the hydrodynamic conditions of the pond leads to the deviations that occur in practice. In this sense,

Table 6.4. Values of K_b for complete mixing, at the temperature of 20 °C, for different values of the depth H, the L/B ratio, and the detention time t, for facultative and maturation ponds

| | | K_b complete mix (d^{-1}) | | | | | | K_b complete mix (d^{-1}) | | | |
| | | L/B ratio | | | | | | L/B ratio | | | |
t (d)	H (m)	1	2	3	4	t (d)	H (m)	1	2	3	4
3	1.0	0.61	0.67	0.72	0.77	20	1.0	1.97	4.34	7.29	10.68
	1.5	0.34	0.36	0.37	0.38		1.5	0.51	0.82	1.19	1.63
	2.0	0.23	0.24	0.24	0.25		2.0	0.42	0.57	0.71	0.84
	2.5	0.17	0.18	0.18	0.18		2.5	0.26	0.33	0.39	0.45
5	1.0	0.72	0.86	0.99	1.12	25	1.0	3.34	7.99	13.76	20.40
	1.5	0.37	0.40	0.43	0.46		1.5	0.69	1.29	2.03	2.88
	2.0	0.24	0.25	0.27	0.28		2.0	0.31	0.45	0.62	0.82
	2.5	0.18	0.18	0.19	0.19		2.5	0.20	0.24	0.30	0.36
10	1.0	1.17	1.67	2.13	2.57	30	1.0	*	*	*	*
	1.5	0.48	0.59	0.70	0.81		1.5	0.95	1.99	3.28	4.76
	2.0	0.28	0.32	0.36	0.40		2.0	0.37	0.62	0.92	1.26
	2.5	0.20	0.21	0.23	0.25		2.5	0.22	0.30	0.39	0.51
15	1.0	1.86	2.90	3.87	4.78	40	1.0	*	*	*	*
	1.5	0.64	0.89	1.11	1.33		1.5	*	*	*	*
	2.0	0.34	0.43	0.51	0.59		2.0	0.57	1.15	1.87	2.69
	2.5	0.22	0.26	0.30	0.34		2.5	0.28	0.47	0.70	0.97

(*) Considerable departure from the validity range of equations 6.6 and 6.7
Shaded cells: more usual values in facultative and maturation ponds
K_b for complete mix: Equations 6.6 and 6.7
K_b for dispersed flow: Equation 6.5
Dispersion number: d = 1/(L/B)

there are the following situations:

- in the *complete-mix* regime, the coefficients obtained experimentally are *larger* than those determined purely according to the kinetics, owing to the fact that the complete-mix reactors are less efficient
- in the *plug-flow* regime, the coefficients obtained experimentally are *smaller* than those obtained purely according to the kinetics, because the plug-flow reactors are more efficient
- in the *dispersed-flow* regime, the coefficients should be *close* to the values according to the kinetics, provided the dispersion number adopted for the pond is correct.

Table 6.4 presents the values of K_b for the complete-mixing hydraulic regime, obtained according to the methodology described above ($K_{b\ disp}$ estimated from Equation 6.5 and $K_{b\ mix}$ estimated from Equations 6.6 or 6.7, according to its applicability range). The values of the dispersion number d were converted to L/B values using Equation 2.14 [d = 1/(L/B)], to make the table more practical. The table presents only L/B ratios up to 4. Higher values could be calculated using equations 6.6 or 6.7 but, for a conceptual point of view, the ideal would be to use the dispersed-flow model, since, in practice, it is known that elongated ponds do not behave as complete-mix reactors.

Table 6.5. Summary of the ranges of typical values of K_b (20 °C) for facultative and maturation ponds, according to the dispersed-flow and complete-mix models

Pond type	Detention time t (d)	Depth H (m)	L/B ratio	K_b dispersed flow (d^{-1})	K_b complete mix (d^{-1})
Facultative	10 to 20 20 to 40	1.5 to 2.0	2 to 4	0.2 to 0.3	0.4 to 1.6 1.6 to 5.0
Maturation (unbaffled, in series)	3 to 5 (in each pond)	0.8 to 1.0	1 to 3	0.4 to 0.7	0.6 to 1.2
Maturation (baffled, single pond)	10 to 20	0.8 to 1.0	6 to 12	0.4 to 0.7	(*)
Maturation (baffled, in series)	3 to 5 (in each pond)	0.8 to 1.0	6 to 12	0.4 to 0.7	(*)

Larger values of K_b: associated to smaller values of t, smaller values of H and larger values of L/B
For values outside the typical ranges: use methodology described in Sections 6.3.4 and 6.3.5
(*) Baffled maturation ponds: adoption of the dispersed-flow model is recommended

6.3.6 Summary of the coliform die-off coefficients

As a summary of all these considerations, Table 6.5 presents the typical range of resultant values of the coefficient K_b, for facultative and maturation ponds, according to the dispersed-flow and complete-mix hydraulic regimes. Values outside the typical ranges may be calculated using the methodologies in Sections 6.3.4 and 6.3.5. It can be observed that the ranges of K_b for dispersed flow are much narrower than those for complete mix, indicating a greater reliability in their estimation.

For other temperatures, K_b can be corrected by the formula:

$$\boxed{K_{bT} = K_{b20}.\theta^{(T-20)}}$$
$$(6.8)$$

where:

θ = temperature coefficient

The values of θ also vary, according to the literature. Very high values ($\theta = 1.19$) were reported by Marais (1974). However, according to Yanez (1993) these values are overestimated, and the values of θ to be adopted should be in the range of **1.07** (7% increase in K_b for an increase of 1 °C in the temperature).

6.4 QUALITY REQUIREMENTS FOR THE EFFLUENT

Normally there are no discharge standards for coliforms. The water quality standards are usually with respect to the receiving body, as a function of its intended uses.

If the effluent is to be used for unrestricted irrigation (for cultures that can present contamination risks), the recommended values according to the World Health Organisation (WHO, 1989) are:

- *faecal coliforms*: $\leq 1{,}000$ faecal coliforms/100 mL (*geometric mean*)
- *helminth eggs*: ≤ 1 egg/L (*arithmetic average*)

For restricted irrigation, there is a limit for only helminth eggs (≤ 1 egg/L), and no limits for coliforms.

In any case, in terms of the receiving body or for agricultural reuse, the coliform counts in the effluent should be very low. Considering that the faecal (thermotolerant) coliform concentrations are in the order from 10^6 to 10^9 org/100mL in the raw sewage, the removal efficiencies in the treatment should be extremely high. To comply with the above criteria, coliform removal efficiencies of the order of 3 to 6 log units (99.9 to 99.9999%) are necessary in the wastewater treatment plant.

It should be noted that the mean referred above for the coliform concentration is expressed in terms of the **geometric mean**. Therefore, it is worthwhile to analyse this statistical parameter. For variables whose *values vary within several orders of magnitude*, it is more convenient to utilise the geometric mean, instead of the arithmetic mean. This is the case in the monitoring of coliforms, which vary within a very wide range, for instance, from 10^6 to 10^9 FC/100mL in raw wastewater. The higher values have a great weight on the arithmetic mean, distorting the concept of the mean as a measure of central tendency. In the range cited, the higher value is 1000 (10^3) greater than the lower value. The calculation of the geometric mean is presented below and illustrated in Example 6.1.

The geometric mean is given by the n root of the product of the n terms:

$$\boxed{\text{Geometric mean} = (x_1 . x_2 \ldots x_n)^{1/n}} \tag{6.9}$$

The geometric mean can be also calculated by:

$$\boxed{\text{Geometric mean} = 10^{\text{(arithmetic mean of the logarithms)}}} \tag{6.10}$$

The following statement is also important, and easily obtainable from the considerations above:

$$\boxed{\text{Log}_{10} \text{ of the geometric mean} = \text{arithmetic mean of the log}_{10}} \tag{6.11}$$

Example 6.1

In a monitoring programme, the following values of faecal (thermotolerant) coliforms have been obtained in four samples: 50, 400, 3000 and 20000 FC/100mL. These data, together with the base-10 logarithms ($\log_{10}$) are presented in the table below.

Coliform data (original data and log transformation)

Data	FC (FC/100 mL)	Log_{10}(FC)
1	5.00E + 01	1.699
2	4.00E + 02	2.602
3	3.00E + 03	3.477
4	2.00E + 04	4.301

Example 6.1 (Continued)

Calculate the geometric and the arithmetic means of the coliform concentrations.

Solution:

Applying Equation 6.9:

$$\text{Geometric mean} = (x_1.x_2.x_3.x_4)^{1/4} = (50 \times 400 \times 3000 \times 20000)^{1/4}$$

$$= 1047 = 1.047 \times 10^3 \, FC/100 \text{ mL}$$

The geometric mean can be also calculated through Equation 6.10. In the example, the arithmetic mean of the $\log_{10}$ of the FC values presented in the table is:

$$\text{Arithmetic mean of the logarithms} = (1.699 + 2.602 + 3.477 + 4.301)/4$$

$$= 3.020$$

Hence:

$$\text{Geometric mean} = 10^{(3.020)} = 1047 = 1.047 \times 10^3 \, FC/100 mL$$

The value found is, of course, equal to the one obtained from Equation 6.9. The calculation using Equation 6.11 leads to:

$$\text{Log}_{10}(1.047) = 3.020$$

If the *arithmetic mean* of the original FC data had been calculated, the following value would have been obtained: 5863 FC/100mL = 5.863 $\times$ 10^3 CF/100mL. This value is much higher than that found through the geometric mean, being greater than 3 from the 4 data available, and not giving, therefore, a good indication of the central tendency of the data.

6.5 DESIGN CRITERIA FOR COLIFORM REMOVAL

The requirement of high efficiencies brings about the need to select a hydraulic regime that allows such high efficiencies. Hence, the maturation ponds should be designed according to one of the following two configurations:

- *baffled pond(s)* (aiming at approaching plug-flow conditions)
- *ponds in series* (preferably three or more)

The main design parameters are: hydraulic detention time (t), pond depth (H), number of ponds (n) and the length/breadth ratio (L/B).

In order to allow a preliminary analysis from the designer with respect to these parameters, Tables 6.6 (temperature of 20 °C) and 6.7 (temperature of 25 °C) present the coliform removal efficiencies that can be obtained in a single pond, for different values of t, H and L/B. The removal efficiencies are reported in terms of logarithmic units removed. The tables were composed according to the methodology proposed for dispersed flow – Equation 6.5 for K_b, Equation 2.14 for d and the formulas in Table 6.1. Table 6.7 was constructed correcting the coefficient K_b for T = 25 °C using the temperature coefficient $\theta = 1.07$. In order to broaden the application of the tables, they include typical depths and detention times, not just for maturation ponds, but also for facultative ponds.

The **overall removal efficiency** in a system comprised by a series of ponds with *different dimensions and characteristics* is given by:

$$E = 1 - [(1 - E_1) \times (1 - E_2) \times \ldots \times (1 - E_n)]$$ (6.12)

where:

E = overall removal efficiency
E_1 = removal efficiency in pond 1
E_2 = removal efficiency in pond 2
E_n = removal efficiency in pond n

In this equation, all removal efficiencies should be expressed as a fraction, and not as percentage (e.g. 0.9, and not 90%).

In case the ponds have the *same dimensions and characteristics*, the formula is simplified to:

$$E = 1 - (1 - E_n)^n$$ (6.13)

where:

E = overall removal efficiency
E_n = removal efficiency in any pond of the series
n = number of ponds in the series

In this equation, all removal efficiencies should be expressed as a fraction, and not as percentage (e.g. 0.9, and not 90%).

If the removal efficiencies are expressed in terms of *log units removed*, the overall removal is given by the sum of the individual efficiencies in each pond, irrespective of the dimensions and characteristics being the same or not:

$$\text{log units} = (\text{log units pond 1}) + (\text{log units pond 2}) + \ldots + (\text{log units pond n})$$

(6.14)

where:

log units = log units removed in the overall system
log units pond 1 = log units removed in pond 1
log units pond 2 = log units removed in pond 2
log units pond n = log units removed in pond n

Waste stabilisation ponds

Table 6.6. Coliform removal efficiencies, expressed in terms of logarithmic units removed, for different values of the hydraulic detention time t, depth H and L/B ratio (dispersed flow). Temperature = **20 °C**

		Log units removed									
		L/B ratio									
t (d)	H (m)	1	2	3	4	6	8	10	12	16	32
3	1.0	0.48	0.51	0.54	0.56	0.59	0.61	0.62	0.63	0.65	0.67
	1.5	0.32	0.34	0.35	0.36	0.38	0.38	0.39	0.39	0.40	0.41
	2.0	0.24	0.25	0.26	0.26	0.27	0.28	0.28	0.28	0.28	0.29
	2.5	0.19	0.20	0.20	0.20	0.21	0.21	0.21	0.21	0.22	0.22
5	1.0	0.68	0.75	0.81	0.85	0.91	0.95	0.97	1.00	1.03	1.09
	1.5	0.48	0.51	0.54	0.56	0.59	0.61	0.62	0.63	0.65	0.67
	2.0	0.36	0.39	0.40	0.41	0.43	0.44	0.45	0.45	0.46	0.47
	2.5	0.29	0.31	0.32	0.32	0.33	0.34	0.35	0.35	0.35	0.36
10	1.0	1.05	1.21	1.33	1.42	1.55	1.65	1.72	1.78	1.87	2.05
	1.5	0.77	0.86	0.92	0.98	1.05	1.10	1.14	1.17	1.21	1.29
	2.0	0.60	0.66	0.70	0.74	0.78	0.81	0.84	0.85	0.88	0.92
	2.5	0.49	0.54	0.56	0.59	0.62	0.64	0.65	0.66	0.68	0.71
15	1.0	1.34	1.57	1.74	1.88	2.08	2.24	2.35	2.45	2.60	2.92
	1.5	0.99	1.13	1.24	1.32	1.44	1.52	1.59	1.64	1.71	1.87
	2.0	0.79	0.89	0.95	1.01	1.09	1.14	1.18	1.21	1.26	1.34
	2.5	0.66	0.72	0.77	0.81	0.87	0.90	0.93	0.95	0.98	1.04
20	1.0	1.57	1.87	2.09	2.27	2.54	2.75	2.91	3.04	3.25	3.72
	1.5	1.17	1.36	1.50	1.61	1.78	1.90	1.99	2.06	2.17	2.41
	2.0	0.95	1.08	1.17	1.25	1.36	1.43	1.49	1.54	1.61	1.75
	2.5	0.79	0.89	0.96	1.01	1.09	1.15	1.19	1.22	1.26	1.35
25	1.0	1.77	2.13	2.40	2.62	2.95	3.21	3.41	3.58	3.85	4.47
	1.5	1.34	1.57	1.74	1.88	2.08	2.24	2.36	2.45	2.60	2.92
	2.0	1.08	1.25	1.37	1.46	1.60	1.71	1.78	1.85	1.94	2.13
	2.5	0.91	1.04	1.13	1.20	1.30	1.37	1.43	1.47	1.53	1.66
30	1.0	1.95	2.37	2.68	2.94	3.33	3.63	3.87	4.08	4.40	5.17
	1.5	1.48	1.76	1.96	2.12	2.37	2.55	2.70	2.82	3.00	3.41
	2.0	1.20	1.40	1.55	1.66	1.83	1.96	2.06	2.13	2.25	2.50
	2.5	1.02	1.17	1.28	1.36	1.49	1.58	1.65	1.71	1.79	1.95
40	1.0	2.27	2.79	3.18	3.50	4.00	4.38	4.70	4.97	5.40	6.46
	1.5	1.73	2.08	2.34	2.55	2.87	3.12	3.32	3.48	3.74	4.32
	2.0	1.42	1.68	1.87	2.02	2.25	2.42	2.55	2.66	2.83	3.20
	2.5	1.21	1.41	1.55	1.67	1.84	1.97	2.07	2.14	2.26	2.52

K_b (dispersed flow) $= 0.542.H^{-1.259}$ $\qquad$ $d = 1/(L/B)$

Log units removed. $= -\log_{10}(1 - \text{Efficiency}/100)$

Efficiency (%) $= 100.(N_o - N)/N_o = 100.(1 - 10^{-\log \text{units removed}})$

Log units removed in a system with ponds in series = sum of the log units removed in each individual pond in the series

Table 6.7. Coliform removal efficiencies, expressed in terms of logarithmic units removed, for different values of the hydraulic detention time t, depth H and L/B ratio (dispersed flow). Temperature = **25°C**

		Log units removed									
		L/B ratio									
t (d)	H (m)	1	2	3	4	6	8	10	12	16	32
3	1.0	0,61	0,66	0,71	0,74	0,79	0,82	0,84	0,86	0,88	0,93
	1.5	0,42	0,45	0,47	0,49	0,51	0,52	0,53	0,54	0,55	0,57
	2.0	0,32	0,33	0,35	0,36	0,37	0,38	0,38	0,39	0,39	0,40
	2.5	0,25	0,26	0,27	0,28	0,29	0,29	0,29	0,30	0,30	0,31
5	1.0	0,85	0,96	1,04	1,10	1,19	1,25	1,29	1,33	1,39	1,49
	1.5	0,61	0,67	0,71	0,74	0,79	0,82	0,84	0,86	0,88	0,93
	2.0	0,47	0,51	0,53	0,55	0,58	0,60	0,61	0,62	0,63	0,66
	2.5	0,38	0,40	0,42	0,43	0,45	0,46	0,47	0,48	0,49	0,50
10	1.0	1,29	1,51	1,67	1,79	1,99	2,13	2,24	2,33	2,47	2,76
	1.5	0,95	1,08	1,18	1,25	1,36	1,44	1,50	1,55	1,62	1,76
	2.0	0,76	0,84	0,91	0,96	1,03	1,08	1,12	1,14	1,18	1,26
	2.5	0,63	0,69	0,74	0,77	0,82	0,85	0,88	0,90	0,92	0,97
15	1.0	1,61	1,93	2,16	2,35	2,63	2,85	3,02	3,16	3,38	3,88
	1.5	1,21	1,41	1,56	1,67	1,84	1,97	2,07	2,15	2,27	2,52
	2.0	0,98	1,11	1,22	1,29	1,41	1,49	1,56	1,61	1,68	1,83
	2.5	0,82	0,92	1,00	1,05	1,14	1,19	1,24	1,27	1,32	1,42
20	1.0	1,88	2,28	2,58	2,82	3,18	3,47	3,70	3,89	4,19	4,90
	1.5	1,43	1,69	1,88	2,03	2,26	2,43	2,57	2,68	2,85	3,22
	2.0	1,16	1,34	1,48	1,59	1,75	1,86	1,95	2,02	2,13	2,36
	2.5	0,98	1,12	1,22	1,30	1,42	1,50	1,56	1,61	1,69	1,84
25	1.0	2,12	2,59	2,95	3,23	3,68	4,02	4,30	4,54	4,92	5,84
	1.5	1,61	1,93	2,16	2,35	2,63	2,85	3,02	3,16	3,38	3,88
	2.0	1,32	1,55	1,71	1,85	2,05	2,20	2,31	2,41	2,55	2,86
	2.5	1,12	1,29	1,42	1,52	1,67	1,78	1,87	1,93	2,03	2,24
30	1.0	2,33	2,87	3,28	3,61	4,13	4,53	4,86	5,14	5,60	6,71
	1.5	1,78	2,15	2,42	2,64	2,97	3,23	3,44	3,61	3,88	4,51
	2.0	1,46	1,73	1,93	2,09	2,33	2,51	2,65	2,77	2,95	3,34
	2.5	1,25	1,45	1,61	1,73	1,91	2,04	2,15	2,23	2,36	2,63
40	1.0	2,70	3,37	3,87	4,28	4,92	5,44	5,86	6,22	6,82	8,32
	1.5	2,07	2,53	2,88	3,15	3,58	3,92	4,19	4,42	4,78	5,66
	2.0	1,71	2,06	2,31	2,51	2,83	3,07	3,26	3,42	3,67	4,24
	2.5	1,47	1,74	1,94	2,10	2,34	2,52	2,66	2,78	2,96	3,36

K_b (dispersed flow) = 0.542.$H^{-1,259}$ $\qquad$ d = 1/ (L/B)

Log units removed. = $-\log_{10}$ (1 − Efficiency/100)

Efficiency (%) = 100. $(N_o - N)/N_o$ = 100.$(1 - 10^{- \log \text{units removed}})$

Log units removed in a system with ponds in series = sum of the log units removed in each individual pond in the series

Regarding the **depth**, *maturation ponds* are usually designed with shallow depths, in order to maximise photosynthesis and the bactericidal effect of the UV radiation. Commonly adopted values are:

$$\boxed{\text{Depth}: H = 0.8 \text{ to } 1.0 \text{ m}}$$

Some researches (von Sperling *et al.*, 2003) have demonstrated the great advantages in terms of efficiency when using ponds with depths lower than 0.8 m. However, the possibility of the growth of rooted plants and the faster filling with sludge are aspects that need to be further investigated.

The introduction of baffles is facilitated due to the low depth of the maturation ponds. The baffles can be built with embankments, wood, pre-cast concrete walls, tarpaulin or plastic membranes supported on structures like internal fences.

When designing the maturation ponds, the previous coliform removal in the upstream units (e.g. anaerobic ponds, anaerobic reactors, facultative ponds) should be taken into consideration. Coliform removal in the facultative ponds can be estimated following the methodology presented in this chapter. For design purposes, the coliform removal in anaerobic ponds or UASB reactors can be adopted as 90% (1 logarithmic unit removed).

Mara (1996) also proposes the observation of the following criterion:

> Minimum detention time in each pond, in order to avoid short circuits and the washing-out of the algae: 3 days

---

Example 6.2

Design a maturation pond system to treat the effluent from a facultative pond (Example 2.3), given the following characteristics:

- Population = 20,000 inhab
- Influent flow = 3,000 m^3/d
- Temperature: T = 23 °C (liquid)
- Faecal (thermotolerant) coliform concentration in the raw wastewater: $N_0 = 5 \times 10^7$ FC/100mL

Data from the facultative ponds (Example 2.3):

- Number of ponds in parallel: 2
- Length of each pond: L = 245 m
- Breadth of each pond: B = 98 m
- Depth: H = 1.8 m
- Hydraulic detention time: t = 28.8 d

Example 6.2 (*Continued*)

Solution:

1. Coliform removal in the facultative ponds

a) Hydraulic regime to be adopted in the calculations

Adopt the dispersed flow regime.

b) Dispersion number d

Adopting Equation 2.14, and knowing that the L/B ratio in each facultative pond is 2.5 (245 m/98 m = 2.5):

$$d = 1/(L/B) = 1/2.5 = 0.40$$

If the formula of Agunwamba (1992) and Yanez (1993) had been used, the values of $d = 0.42$ and $d = 0.37$, respectively, would have been obtained, which are very close to the values obtained above.

c) Coliform removal coefficient

Using Equation 6.5 for dispersed flow, the value of the bacterial decay coefficient is obtained:

$$K_b \text{ (dispersed flow)} = 0.542.H^{-1.259} = 0.542 \times 1.80^{-1.259} = 0.26\ d^{-1}(20\,°C)$$

If Equation 6.4 (based on H and t) had been used, $K_b = 0.18\ d^{-1}$ would have been obtained.
Correcting K_b for 23 °C:

$$K_{bT} = K_{b20}.\theta^{(T-20)} = 0.26 \times 1.07^{(23-20)} = 0.32d^{-1}$$

d) Effluent coliform concentration

Adopting the equation for dispersed flow (Table 6.1), and knowing that the detention time in the facultative ponds is 28.8 days:

$$a = \sqrt{1 + 4K.t.d} = \sqrt{1 + 4 \times 0.32 \times 28.8 \times 0.40} = 3.95$$

$$N = N_0.\frac{4ae^{1/2d}}{(1 + a)^2 e^{a/2d} - (1 - a)^2 e^{-a/2d}}$$

$$= 5.0 \times 10^7.\frac{4 \times 3.95.e^{1/(2\times0.40)}}{(1 + 3.95)^2.e^{3.95/(2\times0.40)} - (1 - 3.95)^2.e^{-3.95/(2\times0.40)}}$$

$$= 8.2 \times 10^5 FC/100mL$$

This effluent concentration from the facultative pond is the influent concentration to the maturation ponds.

Example 6.2 (Continued)

The coliform removal efficiency in the facultative pond is:

$$E = \frac{N_o - N}{N_o} \times 100 = \frac{5.0 \times 10^7 - 8.2 \times 10^5}{5.0 \times 10^7} \times 100 = 98.4\%$$

2. Alternative: three maturation ponds in series

e) Volume of the ponds

Adopt a total detention time equal to 12 days (4 days in each pond).
Volume of each pond:

$$V = t.Q = 4\,d \times 3{,}000\,m^3/d = 12{,}000\,m^3$$

f) Dimension of the ponds

Depth: H = 1.0 m (adopted)

Surface area of each pond: A = V/H = 12,000 m^3/1.0 m = 12,000 m^2
Total surface area: 12,000m^2 × 3 = 36,000 m^2

Dimensions: adopt square ponds (L/B ratio = 1.0) in this example

    Number of ponds: 3
    Length = 110 m
    Breadth = 110 m
    Depth = 1.0 m

Rectangular ponds could have been also adopted, in order to improve the hydraulic characteristics and minimise the dispersion number.

The total area required by the maturation ponds (including banks, roads etc) is around 25% greater than the net area determined. Therefore, the total area required is estimated as 1.25 × 36,000 m^2 = 45,000 m^2 = 4.5 ha (2.25 m^2/inhab.).

g) Coliform concentration in the final effluent

Calculation according to the dispersed flow model:

Dispersion number according to Equation 2.14, for L/B = 1:

$$d = 1/(L/B) = 1/1.0 = 1.0$$

If the formula of Yanez (1993), Equation 2.13, had been applied, a value of d = 0.99 would have been obtained (very close to the value obtained above).
The value of the coliform die-off coefficient is given by (Equation 6.5):

$$K_b \text{ (dispersed flow)} = 0.542.H^{-1.259} = 0.542 \times 1.0^{-1.259} = 0.54\,d^{-1}(20\,^\circ C)$$

If Equation 6.4 (based on H and t) had been used, a value of $K_b = 0.58\ d^{-1}$ would have been obtained.

Example 6.2 (Continued)

For $T = 23\ °C$, the value of K_b is:

$$K_{bT} = K_{b20}.\theta^{(T-20)} = 0.54 \times 1.07^{(23-20)} = 0.66d^{-1}$$

The effluent coliform concentration from the 1st pond in the series is:

$$a = \sqrt{1 + 4K.t.d} = \sqrt{1 + 4 \times 0.66 \times 4.0 \times 1.0} = 3.42$$

$$N = N_o.\frac{4ae^{1/2d}}{(1+a)^2 e^{a/2d} - (1-a)^2 e^{-a/2d}}$$

$$= 8.2 \times 10^5.\frac{4 \times 3.42.e^{1/(2 \times 1.0)}}{(1-3.42)^2.e^{3.42/(2 \times 1.0)} - (1-3.42)^2.e^{-3.42/(2 \times 1.0)}}$$

$$= 1.7 \times 10^5 FC/100mL$$

The removal efficiency in the 1st pond of the series is:

$$E = \frac{N_o - N}{N_o} \times 100 = \frac{8.2 \times 10^5 - 1.7 \times 10^5}{8.2 \times 10^5} = 0.789 = 79\%$$

Considering that the three ponds have the same dimensions, the efficiency of the series of $n = 3$ ponds can be calculated:

$$E_n = 1 - (1 - E_1)^n = 1 - (1 - 0.789)^3 = 0.991 = 99.1\%$$

The coliform concentration in the final effluent is:

$$N = N_0.(1 - E) = 8.2 \times 10^5.(1 - 0.991) = 7.7 \times 10^3\ FC/100mL$$

Calculation according to the complete-mix model:

For illustration and comparison, the calculation for the complete-mix hydraulic regime is presented.

Coefficient K_b (20 °C) for complete mix, based on the coefficient K_b for dispersed flow ($K_b = 0.54\ d^{-1}$, for $T = 20\ °C$), $t = 4.0\ d$ and $d = 1.0$ – according to Equation 6.7:

$$\frac{K_{b\,mix}}{K_{b\,disp}} = 1.0 + \left[0.0540 \times (K_{b\,disp}.t)^{1.8166} \times d^{-0.8426}\right]$$

$$= 1.0 + \left[0.0540 \times (0.54 \times 4.0)^{1.8166} \times 1.0^{-1.4145}\right] = 1.22$$

$$K_{mix} = 1.22 \times K_{disp} = 1.22 \times 0.54 = 0.66\ d^{-1}(20\,°C)$$

For $T = 23\ °C$, K_b is corrected to $K_b = 0.81d^{-1}$.

Example 6.2 (Continued)

The coliform concentration in the final effluent is given directly by the following equation, considering the total detention time of 12 d in all the ponds and the number of ponds n = 3 (see Table 6.1):

$$N = \frac{N_o}{\left(1 + K_b \cdot \frac{t}{n}\right)^n} = \frac{8.2 \times 10^5}{\left(1 + 0.81 \cdot \frac{12}{3}\right)^3} = 1.0 \times 10^4 \, FC/100mL$$

The efficiency of the maturation ponds is:

$$E = \frac{N_o - N}{N_o} \times 100 = \frac{8.2 \times 10^5 - 1.0 \times 10^4}{8.2 \times 10^5} = 0.987 = 98.7\%$$

h) Overall removal efficiency

The *overall* efficiency of the facultative ponds – maturation ponds system in the removal of coliforms is:

- Dispersed-flow model for the maturation ponds:

$$E = \frac{N_o - N}{N_o} \times 100 = \frac{5.0 \times 10^7 - 7.7 \times 10^3}{5.0 \times 10^7} \times 100 = 99.984\%$$

- Complete-mix model for the maturation ponds:

$$E = \frac{N_o - N}{N_o} \times 100 = \frac{5.0 \times 10^7 - 1.0 \times 10^4}{5.0 \times 10^7} \times 100 = 99.980\%$$

$$\text{Log units removed} = -\log(1 - E/100) = -\log(1 - 99.984/100)$$
$$= 3.80 \text{ log units removed}$$

Notes: the dispersed-flow and complete-mix models lead to a global removal efficiency of 99.98% (facultative pond - maturation ponds). The effluent coliform estimations led to: dispersed-flow model: 7.7×10^3 FC/100mL; complete-mix model: 1.0×10^4 CF/100 mL. These deviations are small and should be interpreted taking into account the whole uncertainty in the computations involving coliforms and the rounding-ups made in the calculations.

The proposed system of ponds does not comply with the WHO guidelines for unrestricted irrigation (1×10^3 FC/100 mL), but it can comply with some water body standards, depending on the dilution ratio of the receiving watercourse. In any case, the high contribution given by the maturation ponds in the removal of faecal coliforms can be clearly seen.

If higher removal efficiencies are desired, the total detention time and/or number of ponds can be increased, until the desired effluent quality is reached. In addition, each pond may be more elongated, instead of being square.

However, *the increase in the detention time in each pond must be achieved through the increase in the surface area, and not in the depth. If the depth is*

Example 6.2 (Continued)

increased, the value of K_b will be reduced, and the efficiency will not rise as desired.

If a higher number of ponds in series is adopted, the detention time in each individual pond must be verified to see whether it is greater than or equal to 3 d. For instance, 4 ponds in series, with a total detention time of 12 days will lead to t = 3 days in each pond, which should be the minimum acceptable value, according to Mara (1996).

3. Alternative: Single pond with baffles

j) Volume of the pond

Adopt a detention time equal to 12 days.

Volume of the maturation pond:

$$V = t.Q = 12\,d \times 3{,}000\ m^3/d = 36{,}000\ m^3$$

k) Dimensions of the pond

Depth: H = 1.0 m (adopted)

Surface area: A = V/H = 36,000 m^3/1.0 m = 36,000 m^2

Adopt square external dimensions, but internal dimensions divided with 3 baffles. The baffles can be of tarpaulin, wood, earth banks, or other appropriate material.

External dimensions:

    Length: L = 190 m
    Breadth: B = 190 m

The internal L/B ratio of the pond will be (Equation 6.3):

$$L/B = \frac{L}{B}(n+1)^2 = \frac{190}{190}.(3+1)^2 = 16$$

Due to the division of the internal area with 3 baffles, the pond will have 4 compartments, each one with a length of 190 m and a width of 190/4 = 47.5 m. The pond can be considered as behaving as a rectangular pond, with a L/B ratio = 16, total length L = 190 × 4 = 760 m and width 47.5 m.

The total area required for the maturation pond (including banks, roads, etc.) is around 25% greater than the calculated net area. Therefore, the total area required is estimated as 1.25 × 36,000 m^2 = 45,000 m^2 = 4.5 ha (2.25 m^2/inhab.).

l) Hydraulic regime to be adopted in the calculations

Adopt the dispersed-flow regime.

m) Dispersion number

Adopting Equation 2.14, with L/B = 16:

d = 1/ (L/B) = 1/16 = 0.06

Example 6.2 (Continued)

If the formula of Agunwamba (1992) had been used, the value $d = 0.11$ would have been obtained, along with $d = 0.06$ for the formula of Yanez (1993).

n) Coliform die-off coefficient

The value of the bacterial die-off coefficient can be given by (Equation 6.5):

$$K_b \text{ (dispersed flow)} = 0.542.H^{-1.259} = 0.542 \times 1.0^{-1.259} = 0.54 \text{ d}^{-1} \ (20\,°C)$$

If Equation 6.4 (based on H and t) had been used, a value of $K_b = 0.40 \text{ d}^{-1}$ would have been obtained.

For $T = 23\,°C$, the value of K_b is:

$$K_{bT} = K_{b20}.\,\theta^{(T-20)} = 0.54 \times 1.07^{(23-20)} = 0.66 \text{ d}^{-1}$$

o) Effluent coliform concentration

Adopting the equation for dispersed flow (Table 6.1):

$$a = \sqrt{1 + 4K.t.d} = \sqrt{1 + 4 \times 0.66 \times 12 \times 0.06} = 1.73$$

$$N = N_o.\frac{4ae^{1/2d}}{(1+a)^2 e^{a/2d} - (1-a)^2 e^{-a/2d}}$$

$$= 8.2 \times 10^5.\frac{4 \times 1.73.e^{1/(2 \times 0.06)}}{(1+1.73)^2.e^{1.73/(2 \times 0.06)} - (1-1.73)^2.e^{-1.73/(2 \times 0.06)}}$$

$$= 2.2 \times 10^3 \text{ FC/100 mL}$$

This system also does not comply (although it comes close) with the WHO guidelines for unrestricted irrigation (1×10^3 FC/100 mL), but it can comply with some water body standards, depending on the dilution ratio of the receiving watercourse. In this specific example, the results are slightly better than in the case of the three maturation ponds in series. In any case, the high contribution given by the maturation ponds in the removal of faecal coliforms can be clearly seen.

See comments in item h regarding the improvement in the effluent quality.

p) Removal efficiencies

The efficiency of the maturation pond is:

$$E = \frac{N_o - N}{N_o} \times 100 = \frac{8.2 \times 10^5 - 2.2 \times 10^3}{8.2 \times 10^5} \times 100 = 99.7\%$$

The *overall* efficiency of the facultative ponds – maturation pond systems in the removal of coliforms is:

$$E = \frac{N_o - N}{N_o} \times 100 = \frac{5.0 \times 10^7 - 2.2 \times 10^3}{5.0 \times 10^7} \times 100 = 99.996\%$$

Example 6.2 (Continued)

Log units removed $= -\log(1 - E/100) = -\log(1 - 99.996/100)$
$$= 4.35 \text{ log units removed}$$

Note: If the complete-mix model had been adopted (although it is not indicated for ponds with high L/B ratios), the following results would have been obtained, using the methodology exemplified in item i: $K_{mix}/K_{disp} = 17.84$ (Equation 6.5, with K_b dispersed $= 0.54$ d^{-1} for T $= 20\,°$C); K_b complete-mix $= 0.54 \times 17.84 = 9.67$ d^{-1} (20 $°$C) and K_b complete-mix $= 11.85$ d^{-1} (23 $°$C); effluent FC $= 5.7 \times 10^3$ FC/100 mL. This value of effluent faecal coliforms is close to the value estimated according to the dispersed flow model (2.2×10^3 FC/100mL), indicating the suitability of the proposed approach for the estimation of the effluent coliforms of the ponds. Naturally, priority should be given to the utilisation of the dispersed flow model, due to it being conceptually more adequate.

4. Comparison between the two alternatives

Item	Alternative: 3 maturation ponds in series	Alternative: 1 maturation pond with 3 baffles (4 compartments)
Number of ponds	3 in series	1
Number of baffles	–	3
Total detention time (d)	12	12
Detention time in each pond (d)	4	12
Net area required (ha)	3.6	3.6
Gross area required (ha)	4.5	4.5
Length of each pond (m)	110	190
Width of each pond (m)	110	190
Depth (m)	1.0	1.0
FC in the influent to the facultative pond (FC/100 mL)	5.0×10^7	5.0×10^7
FC in the influent to the maturation pond (FC/100 mL)	8.2×10^5	8.2×10^5
FC in the final effluent (FC/100 mL)	7.7×10^3	2.2×10^3
Efficiency of the maturation ponds (%)	99.1	99.7
Global efficiency (facultative + maturation) (%)	99.984	99.996
Log units removed (global)	3.80	4.35

It can be observed that both alternatives are equivalent from the point of view of land requirements and not so different in terms of the quality of the final effluent. In each alternative, it is still possible to have an optimisation in the design, leading to improvements in the effluent quality. In the selection of the alternative, other items should be investigated, related to costs, topography, soil and other local factors.

Example 6.2 (Continued)

Note: in the calculations, small differences may occur due to rounding errors (the calculations have been done using a spreadsheet, which does not round numeric values).

5. *Arrangement of the ponds (including the facultative ponds)*

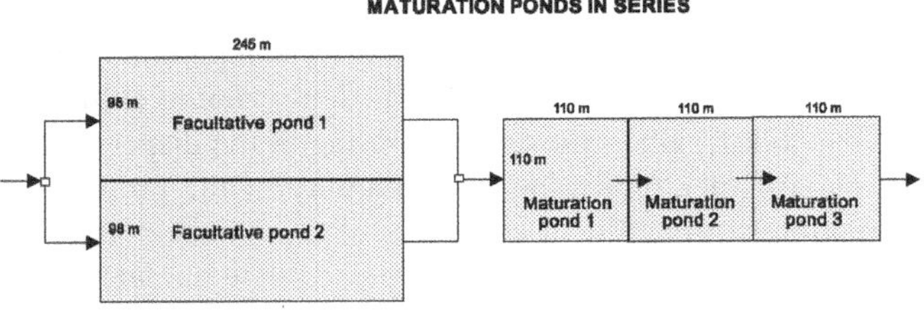

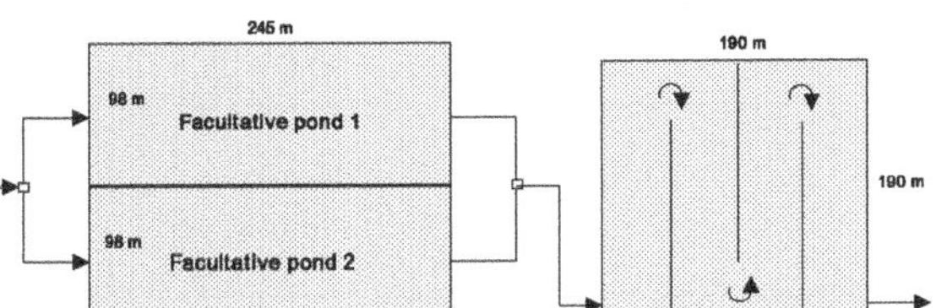

6.6 REMOVAL OF HELMINTH EGGS

6.6.1 Removal of helminth eggs from the wastewater

Helminth eggs are removed by sedimentation, which largely occurs in the anaerobic and facultative ponds. If there are eventually still eggs remaining in the effluent from those ponds, there will be further sedimentation in the maturation ponds. If the WHO guidelines for restricted and unrestricted irrigation (≤ 1 egg/litre) must be satisfied, it can be considered that a system of ponds is likely to produce an effluent that contains frequently zero eggs per litre.

Ayres et al (1992), analysing data of helminth eggs removal in ponds in Brazil, Kenya, and India, developed equations 6.15 and 6.16, valid for anaerobic, facultative and maturation ponds. The equations should be applied sequentially in each pond of the series, so that the number of eggs in the final effluent can be determined (Mara et al, 1992). The model of Ayres et al (1992), applied to a baffled pilot pond in Southeast Brazil, showed good results (von Sperling et al, 2001, 2002).

- Average removal efficiency (to be used to represent average operation conditions):

$$E = 100 \times \left[1 - 0.14.e^{(-0.38.t)} \right] \qquad (6.15)$$

Table 6.8. Removal efficiency of helminth eggs, according to the model of Ayres et al (1992)

Hydraulic detention time (d)	Removal efficiency (%)		Logarithmic units removed	
	Average values	95% confidence	Average values	95% confidence
2	93.45	84.08	1.18	0.80
4	96.94	93.38	1.51	1.18
6	98.57	97.06	1.84	1.53
8	99.33	98.60	2.17	1.85
10	99.69	99.29	2.50	2.15
12	99.85	99.61	2.83	2.41
14	99.93	99.77	3.16	2.64
16	99.97	99.86	3.49	2.85
18	99.985	99.90	3.82	3.02
20	99.993	99.93	4.15	3.17
22	99.997	99.95	4.48	3.28
24	99.998	99.957	4.81	3.37
26	99.999	99.962	5.14	3.42
28	99.9997	99.965	5.47	3.45
30	99.9998	99.964	5.80	3.45

Log units removed $= -\log (1 - E/100)$
Efficiency (%): $E = 100.(1 - 10^{-\log \text{ units removed}})$

- Removal efficiency according to the lower confidence limit of 95% (to be used for design, as a safety measure):

$$E = 100 \times \left[1 - 0.41.e^{(-0.49.t+0.0085.t^2)} \right] \qquad (6.16)$$

where:

E = removal efficiency of helminth eggs (%)
t = hydraulic detention time in each pond of the series (d)

Table 6.8 and Figure 6.7 present the values of the removal efficiency resulting from the application of Equations 6.15 and 6.16.

The concentration to be reached in the effluent also depends largely on the influent concentration. The concentration of eggs in the raw sewage is a function of the sanitary conditions of the population. Typical values are situated in the wide range of **10^1 to 10^3 eggs/L**, with the range between 10^2 and 10^3 eggs/L associated to populations with very unfavourable sanitary conditions. Hence, to reach a final effluent with less than 1 egg/L, for restricted and unrestricted irrigation, the removal efficiencies should be between 90 and 99.9% (1 to 3 log units).

The WHO guidelines specify arithmetic mean values for the helminth eggs. It should be noted, however, that the arithmetic mean is not always the best measure of central tendency, especially in this case, where most of the effluent data have a value of zero, and only a few data have values greater than zero.

Cavalcanti et al (2001) and von Sperling et al (2001, 2002) comment that the removal of helminth eggs is assumed as being a process of discrete settling,

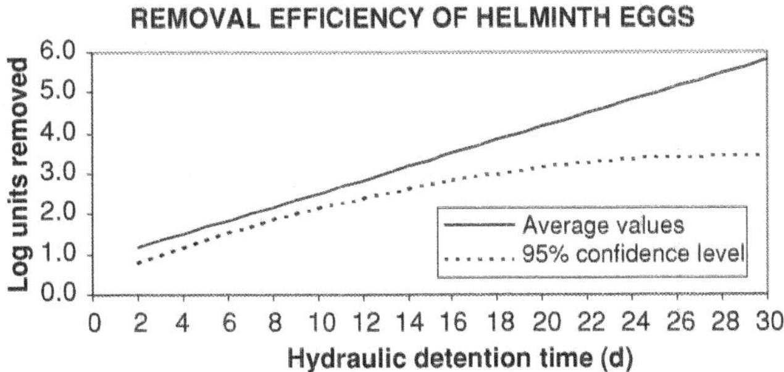

Figure 6.7.   Removal efficiency of helminth eggs, expressed as logarithmic units removed, according to the model of Ayres et al (1992)

which, in theory, is associated with the hydraulic surface loading rate (m^3/m^2.d) and is independent of the depth. Total elimination of helminth eggs was obtained in pilot ponds in Brazil operating with surface loading rates between **0.12** and **0.20 m^3/m^2.d**. The more conservative loading rate of 0.12 m^3/m^2.d with a depth of 1.0 m corresponds to a hydraulic detention time of (1.0 m) / (0.12 m^3/m^2.d) = 8 d.

The WHO (1989) suggests that a series of ponds with total hydraulic detention times of **8 to 10 days** can produce on average effluents with less than 1 egg/litre.

According to Ayres equation (Equation 6.15, for average values), for 8 and 10 days of detention time, the removal efficiency is 2.17 and 2.50 log units (99.3% and 99.7%, respectively). In this case, mean effluent concentrations lower than 1 egg/L will be obtained if the influent has less than 150 to 300 eggs/L.

Figure 6.8 presents the distribution of helminth eggs in the raw wastewater, effluent from a UASB reactor and effluent from the first pond, obtained from five pond systems in Brazil (von Sperling et al, 2003). Some systems had only one pond, while others had ponds in series. It is seen that, already in the effluent from the first pond (or in some cases, the only pond), the eggs concentrations are mostly equal to zero or lower than 1 egg/L. It is worth commenting again that, given the high variability of the data, the *arithmetic mean* is not a good representation of the central tendency, because a few high values tend to increase substantially the average. After the first pond, the *median* values are systematically equal to zero. *Geometric means* may not be calculated, because the occurrence of a single zero value in the whole series leads to a geometric mean of zero, regardless of the other values.

6.6.2 Helminth eggs in the sludge

Research conducted in a baffled pilot pond in Brazil (von Sperling et al, 2002) presented various data of interest with relation to the eggs in the sludge. The settled eggs are incorporated in the bottom sludge, and tend to remain viable for a long period (Figure 6.9). Figure 6.10 presents the longitudinal profile of egg accumulation in the bottom sludge, showing the decreasing tendency along the various compartments of the baffled pond. Also presented are the values of

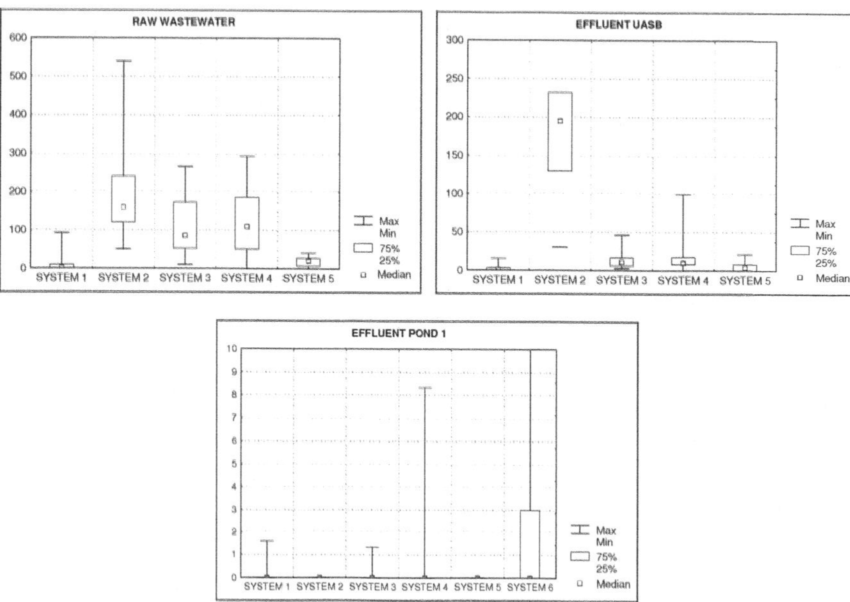

Figure 6.8.   Box-and-whisker plot of the helminth eggs concentrations (eggs/L) in five systems in Brazil, monitored in the following points: influent, effluent from UASB reactors and effluent from first pond

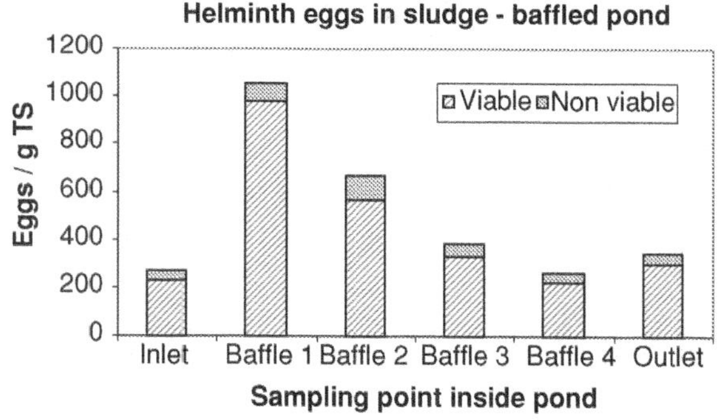

Figure 6.9.   Distribution of helminth eggs in the sludge along a baffled pilot pond in Brazil, after one year of operation, with the indication of the viability and non-viability of the eggs.

the egg counts per gram of total solids, which is a unit usually used for sludge characterisation. Figure 6.11 shows the distribution of the species of helminth eggs in the sludge. It can be observed that the relative distribution is not substantially different along the length of the pond. In terms of the global values in the sludge, the

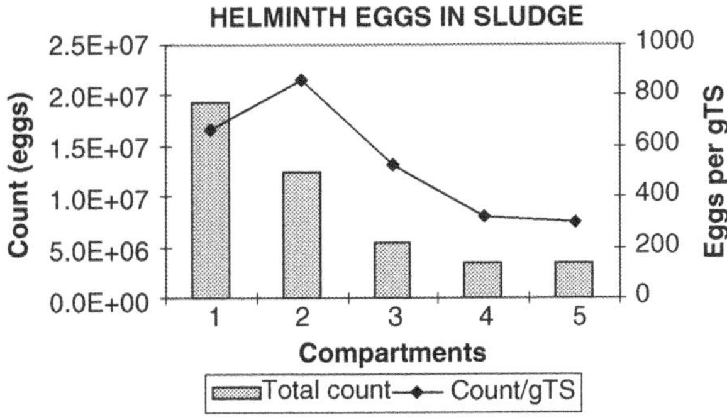

Figure 6.10.   Longitudinal profile of the accumulation of helminth eggs in the sludge in a baffled pilot pond in Brazil, after one year of operation

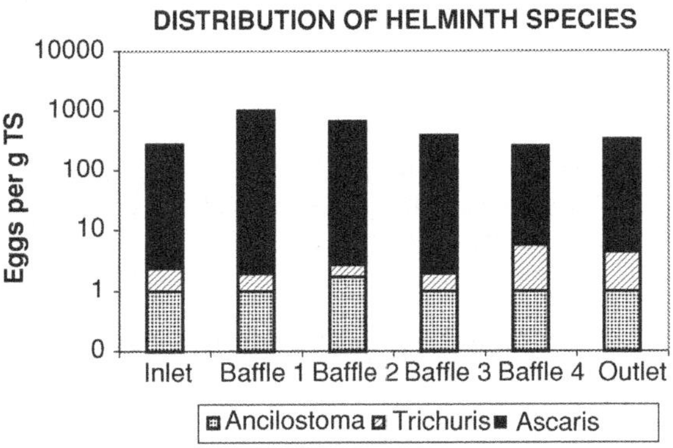

Figure 6.11.   Distribution of the helminth species in the sludge along a baffled pilot pond in Brazil, after one year of operation

following relation was found: *Ascaris lumbricoides*: 99.1%, *Trichuris trichiura*: 0.8%, *Ancilostoma* sp.: 0.1%.

Example 6.3

Estimate the concentration of helminth eggs in the effluent from a system composed of facultative pond – baffled maturation pond (Examples 2.3 and 6.2), with the following characteristics:

Population = 20,000 inhab
Influent flow = 3,000 m^3/d
Concentration of helminth eggs in the raw sewage: 200 eggs/L (assumed)

Example 6.3 (Continued)

Hydraulic detention time in the facultative pond: t = 28.8 d
Hydraulic detention time in the baffled maturation pond: t = 12.0 d

Solution:

a)   Removal of helminth eggs in the facultative pond

For design purposes, the removal efficiency of helminth eggs in the facultative pond is given by Equation 6.16:

$$E = 100 \times \left[1 - 0.41.e^{(-0.49.t+0.0085.t^2)}\right]$$

$$= 100 \times \left[1 - 0.41.e^{(-0.49 \times 28.8 + 0.0085 \times 28.8^2)}\right] = 99.965\%$$

This value is naturally in agreement with the value presented in Table 6.8.

The concentration of eggs in the effluent from the facultative pond is:

$$C_e = C_o \times (1 - E/100) = 200 \times (1 - 99.965/100) = 0.07 \text{eggs/L}$$

The effluent from the facultative pond already complies with the guidelines of the WHO for restricted and unrestricted irrigation (1 egg/L).

b)   Removal of helminth eggs in the maturation pond

Again, for design purposes, the removal efficiency of helminth eggs in the maturation pond is given by Equation 6.16:

$$E = 100 \times \left[1 - 0.41.e^{(-0.49.t+0.0085.t^2)}\right]$$

$$= 100 \times \left[1 - 0.41.e^{(-0.49 \times 12.0 + 0.0085 \times 12.0^2)}\right] = 99.61\%$$

This value is of course the same as that from Table 6.8.

The concentration of eggs in the effluent from the maturation pond (final effluent of the system) is:

$$C_e = C_o \times (1 - E/100) = 0.07 \times (1 - 99.61/100) = 2.7 \times 10^{-3} \text{eggs/L}$$

This value corresponds, in practical terms, to a concentration of zero in the effluent.

7

Nutrient removal in ponds

7.1 NITROGEN REMOVAL

The main mechanisms of *nitrogen* removal in stabilisation ponds are (Arceivala, 1981; EPA, 1983; Soares et al, 1995):

- ammonia stripping
- ammonia assimilation by algae
- nitrate assimilation by algae
- nitrification–denitrification
- sedimentation of the particulate organic nitrogen

Of these mechanisms, the most important is **ammonia stripping**, that is, its release to the atmosphere. In the liquid medium, the ammonia presents itself according to the following equilibrium reaction:

$$\boxed{NH_3 + H^+ \leftrightarrow NH_4{}^+}$$
(7.1)

The free ammonia (NH_3) is susceptible to stripping, while the ionised ammonia cannot be removed by stripping. With the rise of the pH, the equilibrium of the reaction is shifted to the left, favouring the larger presence of NH_3. For 20 °C, in a pH around neutrality, practically all the ammonia is in the form of $NH_4{}^+$. In a pH close to 9.5, approximately 50% of the ammonia are in the form of NH_3 and 50% in the form of $NH_4{}^+$. In a pH greater than 11, practically all the ammonia is in the form of NH_3.

The photosynthesis that takes place in the facultative and maturation ponds contributes to the increase of the pH, through the removal from the liquid of CO_2, that is, carbonic acidity. In conditions of high photosynthetic activity, the pH can rise to values higher than 9.0, providing conditions for the stripping of the NH_3. In addition, under high photosynthetic activity, the high algal production contributes to the direct consumption of NH_3 by the algae (Arceivala, 1981).

The stripping mechanism tends to be more important in *maturation ponds*, which, as a result of their low depths and consequent photosynthetic activity along the whole water column, usually have very high pH values. Additionally, in maturation ponds, the release of oxygen bubbles in the supersaturated liquid phase can accelerate the release of NH_3 (van Haandel and Lettinga, 1994).

In maturation ponds in series, the ammonia removal efficiency can be between 70 and 80%, and in especially shallow maturation ponds it can be greater than 90%, eventually leading to effluent values lower than 5 mg/L of ammonia (van Haandel and Lettinga, 1994; Soares et al, 1995). In facultative and aerated ponds, nitrogen removal efficiency is between 30 and 50%.

The loss of nitrogen through its *assimilation by the algae*, and consequent exit with the effluent is of a smaller importance, in case high removal efficiencies are desired. The nitrogen constitutes around 6 to 12%, in dry weight, of the cellular material of the algae (Arceivala, 1981). Assuming a concentration of 80 mg/L of algae in the effluent, the nitrogen loss will be $0.06 \times 80 \approx 5$ mgN/L to $0.12 \times 80 \approx 10$ mgN/L. Assuming a TKN (ammonia + organic nitrogen) level in the influent in the order of 50 mgN/L, the percentage removal through loss with the final effluent is between 10 and 20%.

The other nitrogen removal mechanisms act simultaneously, but they are considered of less importance. Nitrification is not very representative in facultative and aerated ponds. There is naturally no ammonia oxidation reaction in anaerobic ponds, due to the absence of oxygen.

The literature presents some equations developed in North America for the estimation of the effluent *ammonia* (Equations 7.2 and 7.3) and *nitrogen* (Equations 7.4 and 7.5) concentrations.

Ammonia removal (Pano and Middlebrooks, 1982):

$T < 20\,°C$:

$$C_e = \frac{C_o}{1 + [(A/Q).(0.0038 + 0.000134.T).e^{(1.041+0.044.T).(pH-6.6))}]} \qquad (7.2)$$

$T \geq 20\,°C$:

$$C_e = \frac{C_o}{1 + [5.035 \times 10^{-3}.(A/Q).e^{(1.540\times(pH-6.6))}]} \qquad (7.3)$$

Nitrogen removal (WPCF, 1990; Crites and Tchobanoglous, 2000):
Facultative ponds with a hydraulic regime closer to plug flow:

$$C_e = C_o . e^{\{-K.[t+60.6\times(pH-6.6)]\}} \tag{7.4}$$

$K = 0.0064 \times 1.039^{(T-20)}$

Facultative ponds with a hydraulic regime closer to complete mix:

$$C_e = \frac{C_o}{1 + [t.(0.000576T - 0.00028).e^{(1.08-0.042\times T).(pH-6.6))}]} \tag{7.5}$$

where:

$\quad C_o =$ influent concentration (mg/L)
$\quad C_e =$ effluent concentration (mg/L)
$\quad Q =$ influent flow (m^3/d)
$\quad A =$ surface area of the pond (m^2)
$\quad T =$ temperature of the liquid ($^{\circ}$C)
$\quad pH =$ pH in the pond
$\quad t =$ hydraulic detention time in the pond (d)
$\quad K =$ removal coefficient (d^{-1})

The appropriate equation should be applied sequentially in each pond of the series, in order to lead to the value of the concentration in the final effluent.

Equations 7.2 and 7.3 do not lead to a continuous solution for temperatures lower and greater than 20 °C. The use of Equation 7.2 for values of T close to 20 °C leads to effluent concentration values lower than those from Equation 7.3. Regarding Equation 7.5, it can be observed that it is not very sensitive to variations in the values of pH and T.

The use of the above equations assumes the knowledge of the pH value, a variable that is not known in the design phase. The references above also present the following equation that can be used for the estimation of the pH in the pond, as a function of the alkalinity of the influent sewage:

$$pH = 7.3 \, e^{(0.0005.alk)} \tag{7.6}$$

where:

$\quad$ alk = alkalinity in the influent sewage (mgCaCO$_3$/L)

However, Equation 7.6 does not take into consideration the depth of the pond. It is known (Cavalcanti et al, 2001) that the lower the pond depth, the larger the penetration of the light energy along the water column, photosynthetic activity, consumption of carbonic acidity and rise in the pH. In maturation ponds, pH values higher than those predicted by Equation 7.6 can be reached.

Tables 7.1 and 7.2 and Figures 7.1 and 7.2 respectively present the ammonia and nitrogen removal efficiencies, based on the use of Equations 7.3 and 7.4, for a temperature of 20 °C. For a temperature of 25 °C, Equation 7.3 leads to the same ammonia removal efficiencies, while Equation 7.4 increases the nitrogen

Table 7.1. Ammonia removal efficiency as a function of the Hydraulic Loading Rate (Q/A) and the pH (T $\geq$ 20°C)

Q/A (m³/m².d)	Ammonia removal efficiency (%)				
	pH = 7.0	pH = 7.5	pH = 8.0	pH = 8.5	pH = 9.0
0.025	27	45	63	79	89
0.050	16	29	47	65	80
0.075	11	21	37	56	73
0.100	9	17	30	48	67
0.125	7	14	26	43	62
0.150	6	12	22	39	57

Removal efficiency calculated according to Equation 7.3

Table 7.2. Nitrogen removal efficiency as a function of the hydraulic detention time (t) and the pH (T = 20 °C)

t(d)	Nitrogen removal efficiency (%)				
	pH = 7.0	pH = 7.5	pH = 8.0	pH = 8.5	pH = 9.0
3	16	31	43	53	61
5	17	32	44	54	62
10	20	34	46	55	63
15	22	36	47	57	64
20	25	38	49	58	65
30	29	42	52	61	67
40	34	45	55	63	69

Removal efficiency calculated according to Equation 7.4

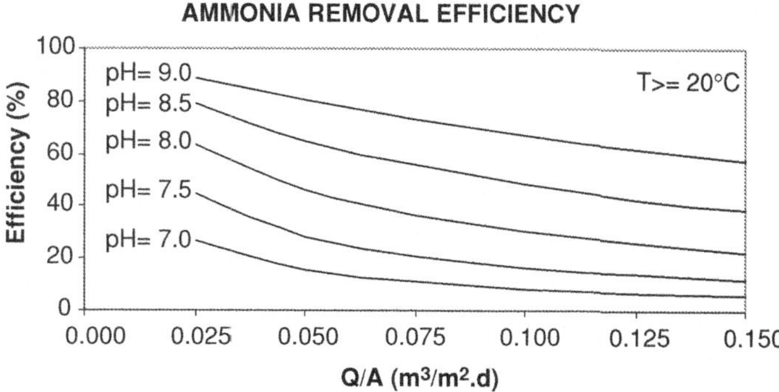

Figure 7.1. Ammonia removal efficiency as a function of the Hydraulic Loading Rate (Q/A) and the pH (T $\geq$ 20 °C) (values from Table 7.1)

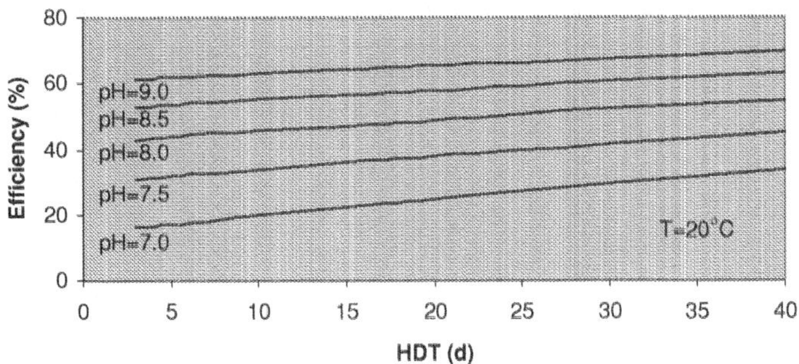

Figure 7.2.   Nitrogen removal efficiency as a function of the hydraulic detention time (t) and the pH (T = 20 °C) (values from Table 7.2)

removal efficiency between 3 and 7%, when compared with the temperature of 20 °C.

Example 7.1

Estimate the ammonia and nitrogen removal in the facultative pond of Example 2.3, whose data are:

- Influent flow: $Q = 3,000 \ m^3/d$
- Surface area: $A = 48,000 \ m^2$
- Hydraulic detention time: $t = 28.8$
- Temperature: $T = 23 \ °C$ (liquid in the coldest month)

The data assumed for the influent are:

- Ammonia = 30 mg/L
- Total nitrogen = 45 mg/L
- Alkalinity: 150 mg/L

Solution:

a)   Ammonia removal

A/Q ratio (reciprocal of the hydraulic loading rate): $A/Q = (48,000 \ m^2) \ / \ (3,000 \ m^3/d) = 16 \ d/m$

(hydraulic loading rate $Q/A = 1/16 = 0.0625 \ m^3/m^2.d$)

pH in the pond (Equation 7.6):

$$pH = 7.3 \ e^{(0.0005.alk)} = 7.3 \ . \ e^{(0.0005 \times 150)} = 7.87$$

Example 7.1 (Continued)

Effluent ammonia concentration (Equation 7.3):

$$C_e = \frac{C_o}{1 + [5.035 \times 10^{-3}.(A/Q).e^{(1.540 \times (pH - 6.6))}]}$$

$$= \frac{30}{1 + [5.035 \times 10^{-3} \times 16 \times e^{(1.540 \times (7.87 - 6.6))}]}$$

$$= 19.1 \, mg/L$$

Ammonia removal efficiency:

$$E = 100 \times (C_o - C_e)/C_o = 100 \times (30 - 19.1)/30 = 36\%$$

This efficiency is in agreement with Table 7.1 and Figure 7.1. The calculations above took into account only the ammonia present in the raw sewage, without considering the fact that a large fraction of the organic nitrogen will be converted into ammonia in the pond itself.

b) Nitrogen removal

Coefficient K:

$$K = 0.0064 \times 1.039^{(T-20)} = 0.0064 \times 1.039^{(23-20)} = 0.0072 \, d^{-1}$$

Effluent nitrogen concentration (Equation 7.4):

$$C_e = C_o.e^{(-K.[t+60.6 \times (pH-6.6)])} = 45 \times e^{\{-0.0072.[28.8+60.6 \times (7.87-6.6)]\}} = 21.0 \, mg/L$$

Nitrogen removal efficiency:

$$E = 100 \times (C_o - C_e)/C_o = 100 \times (45 - 21.0)/45 = 53\%$$

This efficiency is in agreement with Table 7.2 and Figure 7.2.

7.2 PHOSPHORUS REMOVAL

The phosphorus present in the sewage is composed of organic phosphorus and phosphates, with the latter representing the greatest fraction. The main mechanisms of phosphorus removal in stabilisation ponds are (Arceivala, 1981; van Haandel and Lettinga, 1994):

- removal of the *organic phosphorus* contained in the algae and bacteria through its exit with the final effluent
- precipitation of *phosphate* under high pH conditions

The organic phosphorus composes part of the cellular material of the algae. In dry weight, the phosphorus corresponds to values around 1.0% of the algae mass (Arceivala, 1981). Therefore, assuming a concentration of 80 mg/L of algae in the effluent, the phosphorus loss will be around $0.01 \times 80 \approx 0.8$ mgP/L. Admitting a phosphorus concentration in the influent around 8 mgP/L, the percentage removal through loss with the final effluent is only about 10%.

More substantial phosphorus removal can occur through the precipitation of the phosphates under high pH conditions. The phosphates can precipitate in the form of hydroxyapatite or struvite. The same considerations made in Section 7.1 are valid here, emphasising the relation between shallow ponds and high pH values. In the case of phosphorus removal, the dependence of high pH values is larger than with nitrogen: the pH should be at least 9 so that there is a significant phosphorus precipitation. In especially shallow ponds with low hydraulic loading rates, the phosphorus removal is between 60 and 80% (Cavalcanti et al, 2001), while in facultative and aerated ponds, the removal efficiency is usually lower than 35%.

8

Ponds for the post-treatment of the effluent from anaerobic reactors

Anaerobic sewage treatment, and especially the anaerobic sludge blanket reactor (UASB reactor), has grown in popularity and accessibility in many warm-climate countries. Anaerobic reactors reach a good level of efficiency in the removal of BOD (around 60 to 80%), considering the low detention times, the simplicity of the process and the non-existence of equipment, such as aerators. However, this efficiency is most of the time insufficient, bringing about the need for a post-treatment of the anaerobic effluent. The post-treatment can aim at one or some of the following items:

- additional BOD removal
- nutrient removal
- pathogenic organism removal

A very attractive post-treatment alternative is represented by stabilisation ponds, because they maintain in the system the conceptual simplicity already assumed for the anaerobic reactors. This approach of combining anaerobic sludge blanket reactors with stabilisation ponds is believed to have an extremely wide application for developing and warm-climate countries.

The non-mechanised ponds that receive the anaerobic reactor effluent have been designated as **polishing ponds**, to differentiate between the classic concepts of facultative and maturation ponds.

Catunda et al (1994) and Cavalcanti et al (2001) argue that, owing to the BOD removal that takes place in the anaerobic reactors, *the anaerobic effluent can be*

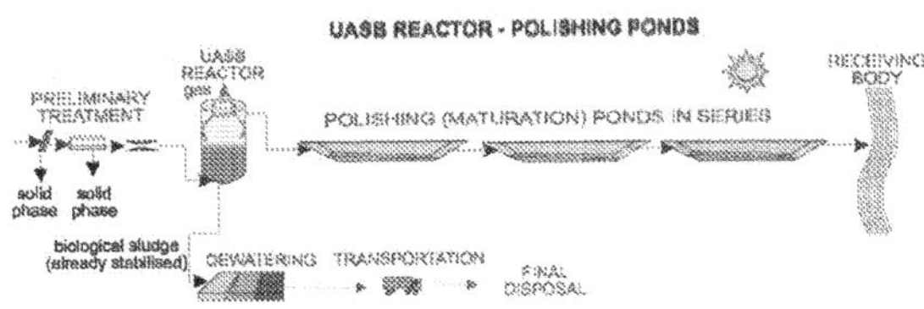

Figure 8.1. Comparison between classical stabilisation pond configurations and the more recent approach of UASB reactors followed by polishing ponds

discharged directly into baffled or in-series polishing ponds, without problems of organic overloading in the first pond of the series or in the initial compartment of the baffled pond. This statement is endorsed by experience acquired in the operation of several polishing ponds in Brazil, as part of PROSAB (Brazilian Research Programme on Basic Sanitation). These pond configurations optimise coliform removal, as commented in Chapter 6. Therefore, the evidence currently available suggests that polishing ponds do not need to be designed as classic facultative ponds, but as maturation ponds (using the design approaches of maturation ponds, regarding the geometric configuration, detention time and depth – see Chapter 6). However, unlike maturation ponds, they not only provide an excellent pathogen removal, but also contribute in a further removal of BOD (hence the name "polishing").

Figure 8.1 presents a comparison of the classical pond configurations with the recent approach of UASB reactors followed by polishing ponds. A significant advantage of this system is the saving in land requirements. It should be understood that the UASB reactor is not simply replacing the anaerobic lagoon, but also the facultative pond.

a) Additional BOD removal

In relation to an additional BOD removal from the anaerobic effluent, this objective can be well accomplished by unaerated ponds or aerated ponds. The first alternative is the most attractive, since it allows a system without mechanisation and with a very low amount of sludge to be treated. The design of the ponds is now for a load of around 20% to 40% of the load of the raw sewage. The land savings are substantial and can make the implementation of ponds possible in locations where mechanised systems would have been previously the only choice. Also in the cases in which the earth movement associated with the construction of a conventional ponds system is excessive, the inclusion of a compact unit such as the anaerobic reactor can contribute to a substantial reduction in the construction costs.

Systems working with this configuration have shown the following characteristics (Cavalcanti et al, 2001):

- absence of mal-odour problems in the ponds (even under high organic load conditions)
- low sludge accumulation in the ponds
- possibility of the use of ponds in series or baffled (without problems of organic overloading in the first pond of the series or in the first compartment of the baffled pond)

The BOD removal coefficients (K) are slightly lower than those of primary facultative ponds, because the stabilisation ponds are already receiving a partially treated influent, in which the easily degradable organic matter has been already removed. However, the coefficients are similar to those used for secondary facultative ponds, following anaerobic ponds.

b) Nutrient removal

Anaerobic treatment systems practically do not remove nutrients. If a high nutrient removal efficiency is required, it should be kept in mind that stabilisation ponds (facultative and aerated) are also not particularly efficient in the removal of N and P. However, polishing (and maturation) ponds can play relatively well this additional role, mainly through the volatilisation of ammonia and the precipitation of phosphates (see Chapter 7).

Ammonia and phosphate removal is greater in polishing ponds with lower depths (less than 1.0 m). In these ponds, the liquid tends to have high pH values, due to the intense photosynthesis that takes place in all the pond volume. The high pH values allow the volatilisation of the free ammonia and the precipitation of the phosphates.

c) Removal of pathogenic organisms

Owing to the low detention times in the anaerobic reactors (in the order of hours), the removal of pathogenic organisms is low in these units (around 1 log unit of coliforms). In this sense, stabilisation ponds, and mainly maturation ponds, can substantially contribute to a high removal efficiency. In the context of post-treatment of anaerobic effluents, the polishing ponds play this role very well, and this is one of their main purposes.

The coliform die-off coefficients are of the same order of magnitude (or maybe slightly higher, owing to receiving a more clarified influent) from those obtained

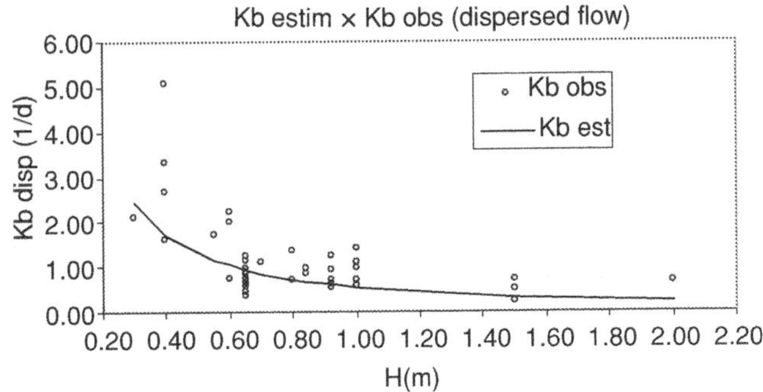

Figure 8.2.   Values of the K_b coefficient (dispersed flow) obtained in 19 polishing ponds in Brazil (n = 45), together with the plot of equation $K_b = 0.542.H^{-1.259}$, based on facultative and maturation ponds in the world (dispersion number d = 1/(L/B))

in facultative and maturation ponds. Figure 8.2 shows K_b values (dispersed flow) obtained in 19 polishing ponds (45 data) in 7 different UASB-polishing pond systems in Brazil (von Sperling et al, 2003b). Equation 6.15 ($K_b = 0.542.H^{-1.259}$), based on 82 facultative and maturation ponds in the world, is also plotted, showing a reasonable fitting to the data. Visually, it is seen that most of the ponds have K_b values slightly higher than those predicted by the overall equation. It can also be seen that shallow ponds (H < 1.0 m) have very high K_b values.

With respect to the removal of helminth eggs, polishing ponds have also been shown to be efficient, similarly to maturation ponds. Effluents with arithmetic means lower than 1 egg per litre are easily achievable and, in most cases, zero counts in the effluent are obtained.

9

Construction of stabilisation ponds

9.1 INTRODUCTION

The operational success of stabilisation ponds depends not only on the process aspects discussed in the previous chapters, but also on the design detailing and on the construction aspects. In general terms, the aspects associated with earthmoving are of fundamental importance, and are likely to have a decisive influence on the economy of the plant. In a more specific aspect, the several details regarding inlet, outlet and interconnection between units are also very important as they have a direct impact on the hydraulic behaviour of the ponds. The detailing aspects should also be considered from the point of view of the operator's needs, in order to make the operational routine of the plant as simple and easy as possible.

This text does not have the objective of furthering the detailing aspects, and other textbooks should be consulted for this purpose. The following topics emphasise just the most important aspects, to which designers and those in charge of the project should give special attention.

9.2 LOCATION OF THE PONDS

The main aspects that should be analysed in selecting the area for the future pond are presented in Table 9.1 (Arceivala, 1981; Silva, 1994).

The simultaneous compliance with the various criteria is usually very difficult, and priority should be given, in each case, to factors of larger importance, which shall be observed according to the local reality.

Table 9.1. Aspects related to the location of the ponds

Aspect	Comment
Area availability	The availability can lead to the selection of the type of pond to be adopted
Location of the area in relation to the wastewater generation location	The closer the pond, the lower the wastewater transportation costs
Location of the area in relation to the receiving body	The closer the pond, the lower the transportation costs of the treated wastewater to the location of its final disposal
Location of the area related to the nearest residences	Anaerobic ponds require a minimum distance of approximately 500 m from the nearest residences, in view of possible bad odours; the other ponds can be located at a shorter distance from the residences
Flood levels	It should be verified whether the land is floodable and the maximum flood levels, for definition of the height of the embankments
Level of the groundwater	The level of the groundwater can determine the settlement level of the ponds and the need to waterproof their bottom
Topography of the area	The topography of the area has a large influence on the earthmoving and, consequently, on the cost of the plant; little sloped areas are preferable
Shape of the area	The shape of the area influences the arrangement of the various units in the floor. In order to save in earthmoving, the shape of the contours can be utilised (provided they are smooth, thus avoiding the creation of dead zones)
Characteristics of the soil	The type of soil has a large influence on the planning of the compensation between cut and fill, on the need of borrow material, on the inclination of the slopes, on the costs of the works (e.g. stones), and on the need of an impermeable bottom
Winds	The location of the pond should permit free wind access, which is important to guarantee a smooth mixing in the pond
Access conditions	Access of the construction teams and of the future operation and maintenance teams should not be difficult
Facility to purchase land	Expropriation difficulties can be an element that affects the feasibility of the pond location in the desired area
Cost of the land	In urban areas or in areas near towns or some important element, the cost of land can be very high, which may lead to the need to adopt more compact treatment processes

With relation to the positioning of the pond in the land, a great effort should be made in the design stage to minimise earthmoving, based on the local topography and geology. Another influencing factor in the location is the direction of the predominant winds. In order to allow a smooth mixing by the wind, *the longest dimension of the pond should be towards the predominant winds*. Should the direction of the winds change seasonally, priority should be given to the direction

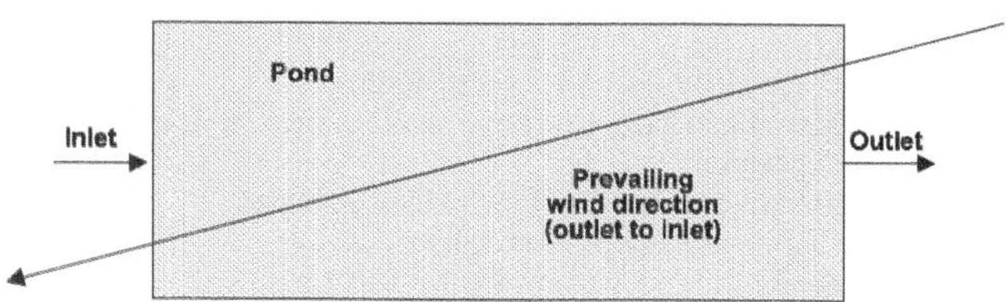

Figure 9.1.   Location of the pond in relation to the direction of the predominant winds

of the wind in the warm period, when thermal stratification is larger. To reduce hydraulic short circuits, *the direction of the wind should be from the outlet to the inlet of the pond* (Mara et al, 1992) (Fig. 9.1).

9.3 DEFORESTATION, CLEANING AND EXCAVATION OF THE SOIL

Deforestation comprises cutting and removal of the trees existing in the area to be occupied by the pond and access roads. The removed material should be moved away from the work site. The cutting of trees shall be approved by the environmental agency.

After the cutting of the trees, the small-sized vegetation is stumped and removed, being, most of the times, burnt in the location itself. The surface of the soil is then raked by motor graders, until the area becomes pure soil (Silva, 1993).

Two different situations may take place in the excavation of the pond (Silva, 1993):

- *Usable excavated material.* This is the desirable situation, in which the cut volumes and the embankment volumes (dikes) are balanced, in order to minimise earthmoving. A large part of the economic feasibility of the construction of a stabilisation pond is associated with the possibilities of earthmoving minimisation. The bottom and the dikes should be compacted in successive earth layers, with humidity and compaction control.
- *Useless excavated material.* This is the case of very sandy soils or those with large quantities of organic matter (peat). In these conditions, the material removed by excavation cannot be used in the construction of the dikes and should be moved away from the location. Borrow material from good quality soil existing near the location should be used for the embankments. After finishing the excavation, the bottom of the pond and the slopes need to be scarified, so that the soil is closely linked to the material used.

9.4 SLOPES

The dikes of the pond are formed by the internal slopes (in contact with the liquid in the pond) and by the external slopes. The aspects listed in Table 9.2 are

Table 9.2. Construction aspects of the pond dikes

Item	Comment
Internal slope	• Usual slope: 1:2 to 1:3 (vert/horiz) • Minimum slope: 1:6 (avoid shallow areas, which allow the growth of vegetation) • Maximum slope: 1:2 (due to land stability) • Clayey soils: slope greater than 1:2 • Sandy soils: slope between 1:3 and 1:6
External slope	• Usual slope: 1:1.5 to 1:2 • Clayey soils: slope greater than 1:2.5 • Sandy soils: slope between 1:5 and 1:8
Slope crest (lane at the crest of the slope)	• Wider than 1.5 m; usually between 2.0 and 4.0 m, in order to allow traffic of the machines during the construction, movement of the maintenance and operation teams, and a possible increase in the height of the dike, if necessary
Freeboard	• Small ponds (<1 ha of area): adopt 0.5 m • Ponds between 1 ha and 3 ha: 0.5 to 1.0 m • Larger ponds: *free board = [log (pond area)]*$^{0.5}$ − 1 (area in m^2) • Purposes: safety in case of increased water level exceeding the design conditions (outlet obstruction, effect of strong winds, new design conception) and safety in case of land settlement due to a contingent lowering of the dike
Waterproofing	• Should the dike material be extremely permeable, it may be necessary to waterproof the dike embankment with clay, geomembranes, sheet piles or concrete slabs • After compaction, the coefficient of permeability should be $<10^{-7}$ m/s
Protection of the internal slopes	• The internal slopes in contact with the water level should be protected against waves, erosion and vegetation growth • The growth of vegetation enables the development of mosquitoes in the ponds (eggs laid in the water and in the shadow of the vegetation) • The types of protection more commonly employed are: large stones (15 to 20 cm), slightly-reinforced concrete slabs (thickness between 7 and 13 cm), concrete plates, reinforced mortar, asphalt pavement, soil cement or plastic membrane • Discontinuous protection (such as stones) enables the growth of vegetation • The protection should extend for at least 0.4 m above and 0.4 m below the water level • Grass or crushed stone should be placed on the rest of the slope over the protection
External slope	• The external slope should be grassed to provide protection against erosion
Corners of the slopes	• The corners of the ponds should be slightly rounded to facilitate the construction and maintenance, and avoid small dead zones
Material of the slopes	• The dikes should be constructed with soil, preferably from the occupied land itself. • The material should be dense, fine, cohesive, and well granulated. • It shall consist of (a) clean soil, without stones and organic matter, and (b) of clay with a little sand
Stormwater drainage	• In ponds that have a side made up of a natural slope (e.g. hill), the stormwater should be collected by ditches parallel to this side, thus preventing the stormwater from passing over the slope

Source: Arceivala (1981), Mara et al (1992), Silva (1993), Jordão and Pessôa (1995)

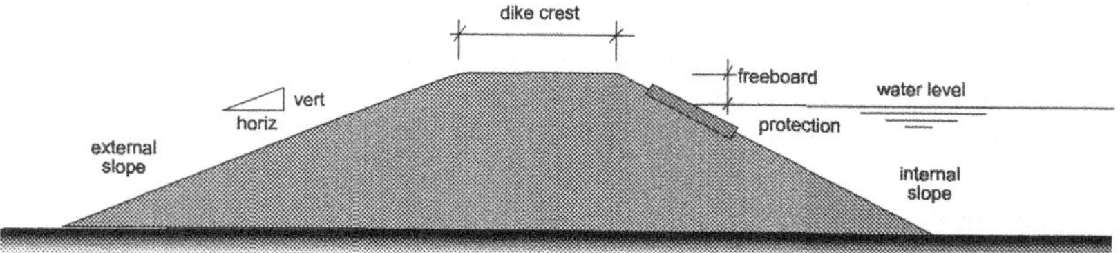

Figure 9.2.   Main elements of a pond dike

important in the construction of the slopes (Silva, 1993; Jordão and Pessôa, 1995). Figure 9.2 lists the main elements of a pond dike.

The length and width determined in the pre-dimensioning are at *mid depth*. The dimensions of the ponds at the bottom, at the water level (WL) and at the crest of the slope depend on the inclination of the internal slope. Assuming that the internal embankment has a slope of 1:d (vertical/horizontal), the referred to dimensions will be:

Length:
- length at the bottom = length at mid depth − 2d.(H/2)
- length at the water level = length at mid depth + 2d.(H/2)
- length at the crest of the slope = length at water level + 2d.(free board)

Width:
- width at the bottom = width at mid depth − 2d.(H/2)
- width at the water level = width at mid depth + 2d.(H/2)
- width at the crest of the slope = width at water level + 2d.(free board)

Example 9.1

Calculate the total dimensions of a pond that has the following dimensions determined in the preliminary design:

- length (at mid depth) = 100.00 m
- width (at mid depth) = 30.00 m
- depth = 2.20 m
- freeboard = 0.60 m
- internal slope = 1:2.5

Solution:

According to the concepts and formulas above:
Internal slope = 1:2.5 → d = 2.5

- length at the bottom = length at mid depth − d.H = 100.00 − 2.5 × 2.20 = 94.50 m
- length at WL = length at mid depth + d.H = 100.0 + 2.5 × 2.20 = 105.50 m

---

Example 9.2 (Continued)

- length at the crest of the slope = length at the WL + 2d.(freeboard) = 105.50 + 2 × 2.5 × 0.60 = 108.50 m

- width at the bottom = width at mid depth – d.H = 30, 00 – 2.5 × 2.20 = 24.50 m
- width at WL = width at mid depth + d.H = 30.0 + 2.5 × 2.20 = 35.50 m
- width at the crest of the slope = width at WL + 2d.(freeboard) = 35.50 + 2 × 2.5 × 0.60 = 38.50 m

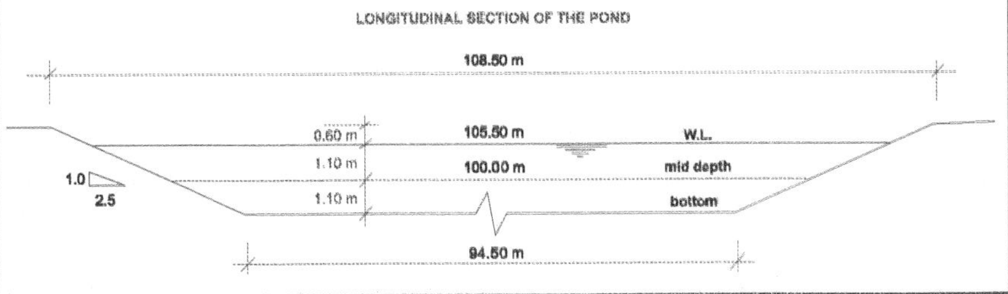

9.5 BOTTOM OF THE PONDS

The bottom of the stabilisation ponds should not lead to excessive seepage, which could cause one of the following problems:

- contamination of the groundwater
- difficulty in maintaining the liquid level in the ponds

The permeability of the soil and the possible interference with the groundwater should be investigated by means of boreholes. It is worth to mention that the sites usually available for possible construction of wastewater treatment plants are frequently located in swamps, marshy areas or with a high groundwater level. The permeability of the bottom tends to decrease as time goes by, as a result of the clogging caused by solids from the sewage and by the biomass. According to Arceivala (1981), under favourable conditions, the losses by infiltration amount to less than 10% of the flow from the pond, being frequently lower than 1%.

Mara et al (1992) propose the following interpretations of the coefficient of permeability k:

- $k > 10^{-6}$ m/s: the soil is very permeable and the bottom should be protected
- $k > 10^{-7}$ m/s: some infiltration may occur, but not enough to prevent the pond from being filled
- $k < 10^{-8}$ m/s: the bottom of the pond will be naturally sealed
- $k < 10^{-9}$ m/s: there is no risk of contamination
- $k > 10^{-9}$ m/s: if the groundwater is used for domestic supply, hydrogeological studies should be performed

A *reduced percolation rate* can be achieved by means of a well-compacted 5–10 cm thick homogeneous clay layer. The *waterproofing* of the bottom can be accomplished through (Jordão and Pessôa, 1995):

- clay layer, with a minimum thickness of 40 cm
- asphalt coating
- plastic geomembranes

The solution to be adopted will naturally have a great impact on the total cost of the pond and on its own economic feasibility.

Should clay be adopted, after preparing the bottom, one should not wait too long for the filling of the pond with liquid (although partial, 1/3 of the height), in order to prevent drying and cracking of the bottom layer.

When estimating the required clay volume, it should be remembered that the clay, after compaction, has its volume reduced. Thus, the clay volume to be acquired should be greater than the volume of the bottom layer.

In aerated ponds, the placement of a concrete plate underneath each aerator is necessary in order to avoid erosion problems caused by the turbulence generated by the aerator.

The pond bottom should be made as level as possible, unless there is an specifically designed hopper near the inlet to retain settleable solids. The finished elevation should not vary more than 15 cm from the average elevation of the bottom (Metcalf & Eddy, 1991).

9.6 INLET DEVICES

The influent wastewater should undergo a preliminary treatment consisting of:

- *Screen.* The screens are usually manually cleaned in most of the ponds. The adoption of mechanised screens is justifiable in ponds of large dimensions or in special situations.
- *Grit chamber.* Although the amount of sand is relatively small compared with all the sludge volume accumulated at the bottom of the pond, the sand tends to settle close to the inlet, which may cause localised problems. As the grit chamber is a small unit of easy maintenance, its inclusion is recommended in all pond systems.
- *Flume or weir for flow measurement.* The flow measurement is essential for the operational control of the pond. The flume also carries out the function of regulating the velocity in rectangular grit chambers. A convenient location for the collection of samples of the influent to the pond is close to the flow measurement unit.

The inlet to the pond should meet the following conditions:

- guarantee a broad homogenisation of the liquid, avoiding the occurrence of hydraulic short-circuits and dead zones
- be submerged, in order to avoid the release of malodorous gases
- avoid erosion of the slopes and the pond bottom (for this purpose, a concrete plate is placed on the bottom, underneath the pipe discharge)

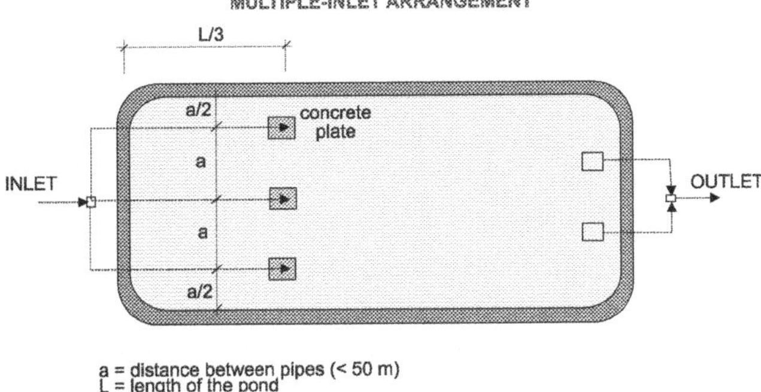

Figure 9.3.   Distribution of the inlet pipes in a pond. Alternative of multiple inlets and outlets in wide ponds.

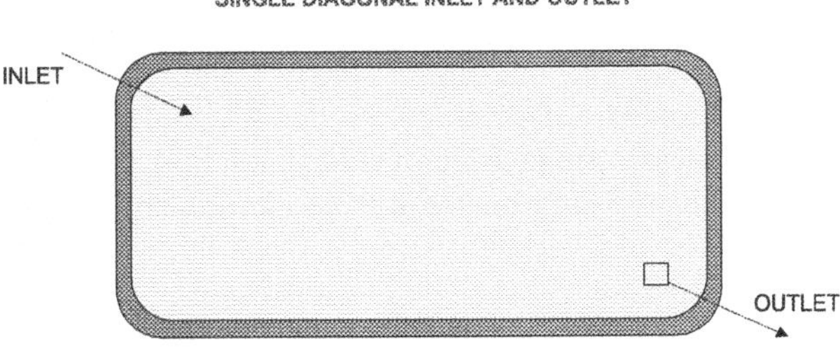

Figure 9.4.   Distribution of the inlet and outlet pipes in a pond. Alternative of single inlets and outlets, located in diagonally opposite ends.

With relation to the number of inlet pipes, there are two approaches in the literature. One states that the homogeneous distribution of wastewater along the width of the pond should be guaranteed through the placement of a sufficient number of inlet pipes (**multiple inlets**). Only small ponds should have a single inlet. Larger ponds should have two, three, or more inlets, distant 50 m at the most one from the other (see Figure 9.3). The inlet should not be located in front of the outlet of the pond, even if at long distances, as this facilitates the occurrence of hydraulic short circuits.

Another approach (Mara et al, 1992) suggests, for simplicity, **single inlets and outlets** in each pond, located diagonally in opposite ends. The argument is that in the case of multiple inlet and outlet structures there may be differential settlements in the structures, altering the relative distribution of the flows (Fig. 9.4).

Should a hydraulic regime approaching that of *complete mix* be desired in the pond, the inlet pipes should extend to *1/4 to 1/3 of the length* of the pond (Figure 9.3). Should an approximation to a *plug-flow* reactor be desired, the inlet pipes should discharge *closer to the inlet side*. It is worth reminding that, for primary ponds, plug-flow conditions should be avoided as they can cause organic overload close to the inlet end. However, for single-cell maturation ponds, the plug-flow system is thoroughly more advantageous, and no overloading problems at the inlet end are expected.

The inlet tubing should be designed for an average flow velocity equal to or higher than 0.5 m/s (Silva, 1993).

In deeper ponds, dead zones are more likely to occur. In these ponds, the design of the inlets and outlets should be made very carefully. In anaerobic ponds, there are indications that a homogeneous distribution at the bottom by means of perforated laterals can contribute to a larger contact between the wastewater and the biomass, thus increasing the efficiency of the pond.

Figure 9.5 presents some types of inlet devices commonly used by designers (Jordão and Pessôa, 1995). Inlets right at the bottom may suffer blocking problems due to localised silting, in case grit removal is inefficient or the wastewater collection system receives large portions of stormwater, which may include substantial loads of sand.

By-pass tubing should be included, allowing the start-up of the facultative pond prior to the anaerobic pond, the interruption in the feeding to a certain pond during sludge removal or for any other operational or maintenance reason.

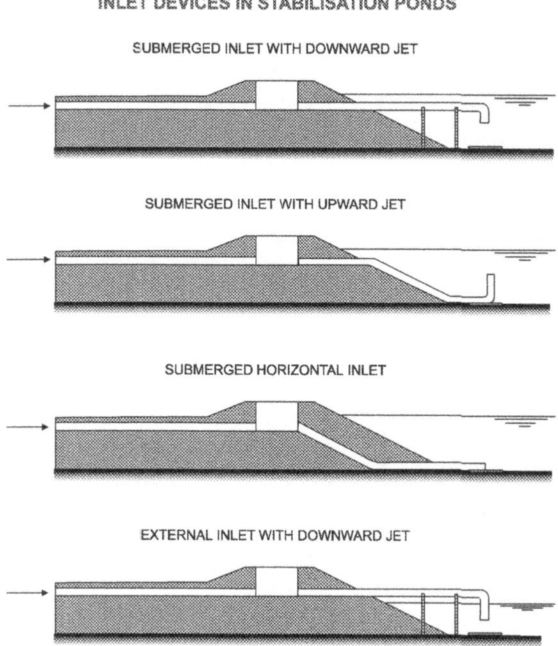

Figure 9.5.   Different inlet schemes in stabilisation ponds (adapted from Jordão and Pessôa, 1995)

9.7 OUTLET DEVICES

The design of the effluent outlet of stabilisation ponds should take into consideration the following aspects (Mara et al, 1992; Silva, 1993; Jordão and Pessôa, 1995):

- the outlet should be located at the opposite end to the inlet, to avoid short circuits
- the outlet should not be aligned with the inlet, in order to minimise short circuits
- the outlet devices can be of either fixed or variable level (the latter is preferable as it allows more flexibility)
- the outlet should have baffles reaching below the water level, to prevent the exit of floating material, such as algae, in the facultative ponds, or scum, in the anaerobic ponds

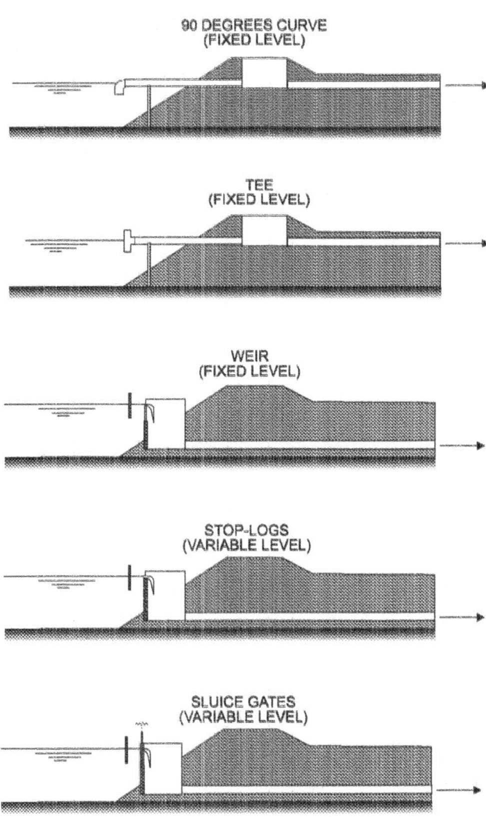

Figure 9.6.   Some outlet devices for stabilisation ponds

- the effluent removal level, dictated by the baffles, has the following conflicting results: shallower removal – higher DO and SS contents; deeper removal – lower DO and SS contents
- approximate values for the effluent removal level shall have the following depths below the water level: anaerobic ponds: 0.30 m; facultative ponds: 0.60 m; maturation ponds: 0.05 m
- the design should allow operational flexibility to adjust the removal level below the water level, in order to achieve the desired point regarding the conflicting objectives between the DO and SS concentrations
- access to the outlet device should be easy, in order to allow flow measurements, collection of samples and changes in the pond water level
- a bottom discharge system can be adopted in the outlet structure itself (although its infrequent or unlikely use may make its utilisation difficult, after a long time of closure)

There are several types of outlet devices. Figure 9.6 illustrates some of these types, such as: (a) fixed water level: *90°curve*, cast iron *tee*, *weir* and (b) variable water level: wooden *stop-logs* (the placement or removal of the wooden boards allows variation in the level of the pond) and *sluice gates*. Stop-logs do not provide a good watertightness, and liquid may pass between the boards, although, in principle, this does not represent a great problem. Screw-driven sluice gates can be adopted, which allow variation of the water level, with a greater watertightness (although not complete).

When the effluent from a pond goes to another pond downstream, the interconnection between the ponds should include an open visiting *box*, in order to enable the collection of samples and the unblocking of pipes.

10

Maintenance and operation of stabilisation ponds

10.1 INTRODUCTION

The conceptual simplicity of stabilisation ponds brings as a consequence the simplicity of the operation and maintenance procedures. The ponds are inherently simple, and they should be designed to be so in their operational routine. In this simplicity lies the great sustainability of wastewater treatment by stabilisation ponds, mainly in developing countries. However, the operational simplicity should not be an excuse for a lack of care with the plant and the process. There are several operation and maintenance procedures that should be carried out following a certain routine, without which environmental problems and a reduced treatment efficiency treatment will take place.

The present chapter deals with the following aspects related to the operation and maintenance of the ponds:

- dimensioning of the operational staff
- inspection and monitoring
- start-up of operation
- operational problems

The coverage of these items in the book is very simple. The references WEF (1990), Mara et al (1992), Yanez (1993) or Jordão and Pessôa (1995) should be consulted for further details with relation to these topics.

It is essential that the design of the stabilisation pond includes an Operation Manual, providing the main guidelines for the suitable operation of the designed system. During the plant operation, the operator can seek the optimisation of the process, based on his accumulated experience in the pond at issue.

10.2 OPERATIONAL STAFF

In a stabilisation pond, most of the personnel are associated with simple maintenance activities, such as grass cutting, cleaning and others. The need for qualified technical personnel is low, compared with most of the other treatment processes. Table 10.1 presents some suggestions for structuring the operational staff with different pond sizes.

10.3 INSPECTION, SAMPLING AND MEASUREMENTS

The operator should carry out a daily inspection throughout the pond and its complementary units. Table 10.2 shows an example of an inspection checklist.

The sampling and measurement scheduling can follow the model presented in Table 10.3. Certainly, depending on the size and importance of the pond, the number of parameters to be included, as well as the frequency of their determination, can be altered and adapted to local needs. Small-sized ponds in remote and lower-income locations can naturally have a more simplified sampling scheduling, concentrated on the determination of the flow and parameters set forth by the environmental legislation. Should effluents be reused in agriculture, agronomic (electric conductivity, Ca, Mg, Na, boron and others) and sanitary (helminth eggs) parameters of interest should be investigated.

Owing to the daily variation of several constituents in stabilisation ponds, composite sampling is preferable. The portions that constitute the composite sample are collected either automatically (automatic samplers) or manually, at 1–3-hour intervals. Should there be any difficulty to collect the composite samples, the collection of a single sample from the water column in the pond leads to results comparable with those of the composite sample. Mara et al (1992) present details of the column sampler.

An aspect of fundamental importance in a monitoring programme relates to the real use of the data surveyed. There is no sense in obtaining data if they are not checked and interpreted. Pond performance monitoring graphs should be produced, with participation of the operator. Data input in computer spreadsheets in the head office, including loading rates, efficiency parameters and associated graphs, is the best form to use these data.

10.4 OPERATION START-UP

10.4.1 Loading of the ponds

The initial loading of the ponds can be done by means of one of the two procedures described below (CETESB, 1989). The loading should be performed preferably in summer, when temperatures are higher.

Table 10.1. Dimensioning of the staff for administration, operation, and maintenance of stabilisation ponds, as a function of the population served

Personnel	Pop ≤ 10,000 inhab.		Pop = 20,000 to 50,000 inhab.		Pop > 50,000 inhab.	
	Facultative pond	Aerated pond	Facultative pond	Aerated pond	Facultative pond	Aerated pond
Administration						
Superintendent engineer	–	–	1/2	1/2	1	1
Secretary	–	–	1/2	1/2	1	1
Assistant	–	–	1	1	1	1
Driver	–	–	1	1	1	1
Operation/maintenance						
Head engineer	1/4	1/4	1/2	1/2	1	1
Chemist	–	–	1/4	1/4	1/2	1/2
Laboratory technician	–	–	1/2	1/2	1	1
Mechanical–electrician	–	–	–	1/2	–	1
Operator- 08:00–16:00 h	1	1	1	1	1	1
Operator 16:00–24:00 h	–	–	–	1	1	1
Operator 24:00–08:00 h	–	–	–	1	–	1
Hand labourers	2	2	2–5	2–7	6–10	7–12

Source: adapted from Yanez (1993)

Table 10.2. Aspects to be included in a daily inspection and occurrence record

Day			
Weather conditions			
weather (sunny, cloudy, rainy) *wind* (none, weak, strong)			
Item	Yes	No	Comment / location / quantity / measures
Observations in the pond			
Is there sludge rising in the pond?			
Are there green patches on the surface?			
Are there black patches on the surface?			
Are there oil stains on the surface?			
Is there vegetation in contact with the water?			
Is there erosion on the slopes?			
Is there visible seepage?			
Are birds present?			
Are insects present?			
Other aspects			
Are the fences in good condition?			
Are the stormwater ditches clean?			
Is the flow meter working?			
Have weeds been removed?			
Has scum been removed?			
Have solids been removed from the screen?			
Has grit been removed from grit chamber?			
Has there been any power failure?			
Has the by-pass to the receiving body been used?			

Source: adapted from Jordão and Pessôa (1995) and Soares (1995)

a)   *Filling of the pond with water pumped from a neighbouring stream or from a public supply system*

- Fill the pond with a minimum water depth, preferably reaching 1 m.
- Close the outlet devices.
- Begin the introduction of sewage until reaching the water depth adopted in the design.

The adoption of this procedure:

- prevents the uncontrolled growth of vegetation, which occurs in conditions of low water depth;
- allows testing of the watertightness of the system;
- enables the correction of occasional problems resulting from a deficient compaction (before the introduction of sewage).

b)   *Filling of the pond with a mixture of water pumped from the stream and wastewater to be treated*

- Mix the wastewater and the water (dilution at a ratio equal to or greater than 1/5)

Table 10.3. Measurement and sampling programme in stabilisation ponds

Frequency	Parameter	Measurem. / analysis	Influent	Facultative pond	Aerated lagoon	Effluent
Daily	Flow (m^3/d)	*In situ*	x			x
	Air temperature (°C)	*In situ*				
	Liquid temperature (°C)	*In situ*	x	x	x	x
	pH	*In situ*	x	x	x	x
	Settleable solids (mL/L)	*In situ*	x			x
	Dissolved oxygen (mg/L)	*In situ*		x	x	
Weekly	Total BOD (mg/l)	Central lab.	x			x
	Total COD(mg/l)	Central lab.	x			x
	Filtered BOD or COD (mg/L)	Central lab.				x
	Faecal coliforms (or *E. coli*) (org/100 mL)	Central lab.	x			x
	Total suspended solids (mg/L)	Central lab.	x			x
	Volatile suspended solids (mg/L)	Central lab.	x			x
Monthly	Organic nitrogen (mg/L)	Central lab.	x			x
	Ammonia (mg/L)	Central lab.	x			x
	Nitrate (mg/L)	Central lab.				x
	Phosphorus (mg/L)	Central lab.	x			x
	Sulphate (mg/L)	Central lab.	x			x
	Sulphide (mg/L)	Central lab.	x			x
	Alkalinity (mg/L)	Central lab.	x			
	Oils and greases (mg/L)	Central lab.	x			x
Occasional	Counting of zooplankton	Central lab.		x		
	Counting of phytoplankton	Central lab.		x		
	Main genera of algae	Central lab.		x		
	DO produced by photosynthesis (mg/L.h)	*In situ*		x		
	DO consumed by respiration (mg/L.h)	*In situ*		x		
	Hourly flow (m^3/h) (24h,every h)	*In situ*	x			
	Hourly DO (mg/L) (24h,every h)	*In situ*		x	x	

Notes: *Anaerobic ponds* – the programme can be similar to that of facultative ponds (including raw sewage), excluding the determination of algae genera, DO production studies, and light penetration; *Maturation ponds* – the programme can be similar to that of facultative ponds; *Use of effluent in agriculture* – include agronomic (electric conductivity, Ca, Mg, Na, B and others) and sanitary (helminth eggs) parameters
Source: adapted from CETESB (1989), WEF (1990), Yanez (1993), Silva (1993), Jordão and Pessôa (1995)

- Fill up the pond to a depth of approximately 0.40 m
- Await some days, until the appearance of algae is visible
- In the subsequent days, add more wastewater, or wastewater/water mixture, until algal blooming occurs
- Interrupt feeding for a period of 7 to 14 days
- Fill up the pond with wastewater until the operation level
- Interrupt the feeding
- Await the establishment of a population of algae (around 7 to 14 days)
- Feed the pond normally with the wastewater

Should no water be available, the ponds can be filled up with raw sewage and left for about 3 to 4 weeks, in order to allow the development of the microbial population. Some odour release will be unavoidable in this period (Mara et al, 1992).

The whole loading period should be monitored by operators with experience in the process. The total loading period can last 60 days, until a balanced biological community is established in the medium.

The following two procedures should be avoided:

- *Feed with the wastewater load adopted in the design, but without a balanced biological community established in the pond*. If this happens, the pond will suffer from anaerobiosis, with release of bad odours. The reversal of the anaerobiosis process can take two months.
- *Feed the ponds with small, continued loads*, which frequently occur when there are few housing connections. In this case, as the soil is not clogged yet, the liquid could percolate through the slopes, accumulating decomposable solids and releasing bad odours.

10.4.2 Beginning of operation of anaerobic ponds

The beginning of the operation of anaerobic ponds requires the following procedures (CETESB, 1989):

- Begin the introduction of sewage according to the recommendations in Section 10.4.1.
- Maintain the pH of the medium slightly alkaline (7.2 to 7.5). To facilitate the occurrence of these conditions, digested sludge from sewage treatment plants or from Imhoff tanks, or limestone, vegetable ash or sodium bicarbonate can be added after 30 days of operation.

Anaerobic ponds should be started-up after the facultative ponds. This avoids the release of odours from the discharge of anaerobic effluents into an empty facultative pond. Should the concentration of raw sewage be very low, or its flow be small in the beginning of the operation of the system, it would be better to divert the raw sewage to the facultative pond, until a volumetric organic load of at least 0.1 kgBOD/m^3.d is reached in the anaerobic pond (Mara et al, 1992). If there is more than one anaerobic pond in parallel, only one pond could be loaded, so that the load applied to this pond is the same as or higher than the minimum value of 0.1 kgBOD/m^3.d.

10.4.3 Beginning of operation of facultative ponds

The following procedures are recommended (CETESB, 1989):

- Begin the introduction of sewage according to the recommendations of Section 10.4.1.
- The maintenance of a slightly alkaline pH should happen naturally, in case the recommendations of Section 10.4.1 are followed.
- Measure the dissolved oxygen daily.

10.4.4 Beginning of operation of ponds-in-series systems

The ponds located downstream of the primary pond can be started-up according to the following recommendations (CETESB, 1989):

- Begin the filling of the ponds when the water depth in the primary pond reaches a minimum value of 1.0 m.
- Close the outlet devices of the ponds.
- Water should be added to the ponds until a depth of 1.0 m is reached.
- When the primary pond reaches the operational level, its effluent can be directed to the subsequent cell, taking the following precautions:
    - Remove the stop-logs slowly, preventing the water depth of the previous unit from dropping below 1.0 m
    - Do not perform bottom discharge operations from the primary cell
    - Equalise the water depth in all ponds slowly
    - Avoid the situation in which a pond is totally full, while the subsequent unit is empty

10.5 OPERATIONAL PROBLEMS

The main operational problems of anaerobic, facultative and aerated ponds are listed in Tables 10.4, 10.5 and 10.6, together with the main measures to be taken for their possible solution (source: CETESB, 1989; WEF, 1990; Silva, 1995; Jordão e Pessôa, 1995).

Table 10.4. Main operational problems of **anaerobic ponds** and their possible solutions

Problem: bad odours

Causes

- Sewage overload and small detention time
- Very low load and an excessively high detention time (the pond behaves as a facultative one, with the presence of DO in the liquid)
- Presence of toxic substances
- Abrupt fall of the wastewater temperature

Prevention and control measures

- Recirculate the effluent from the facultative or maturation pond to the inlet of the anaerobic pond (recirculation ratio of approximately 1/6)
- Improve the distribution of the influent to the pond (distribution by perforated tubes on the bottom of the pond)
- In case of overload apply occasional partial by-pass to the facultative pond (if it supports the increased load)
- In the case of long detention times, operate with only one anaerobic pond (if there are two or more ponds in parallel)
- Add sodium nitrate to several points of the pond
- Add lime ($\sim$12 g/m^3 of the pond) to raise the pH, reducing the acid conditions responsible for the inhibition of methanogenic organisms and for the larger presence of sulphide in the free, toxic form
- Add products that remove sulphides
- Avoid the addition of chlorine, because it will cause subsequent problems to the restart of the biological activities

Problem: proliferation of insects

Causes

- Screened material or sand removed not conveniently disposed of
- Growth of vegetation where the water level is in contact with the internal slope
- Oil and scum layer always present
- Poor maintenance

Prevention and control measures

- Bury the material removed from the screens and grit chambers
- Cut the grown vegetation
- Revolve, with a rake or water jet, the layer of floating material that covers the ponds
- Apply carefully insecticides or larvicides to the scum layer

Problem: growth of vegetation

Causes

- Inadequate maintenance

Prevention and control measures

- Aquatic vegetation (that grows on internal slopes): total removal, preventing it from falling in the pond
- Terrestrial vegetation (that grows on external slopes): remove weeds from the soil; add chemical products for control of weeds

Problem: green patches where the water level is in contact with the slope

Causes

- Proliferation of algae, in view of the small depth in the water level-slope section

Prevention and control measures

- Remove the algae colonies

Problem: blocking of the inlet pipes

Causes

- Inlet pipes obstructed

Prevention and control measures

- Clean the pipes with a stick or steel wire

Problem: surface of the pond covered with a scum layer

Causes

- Scum, oils and plastics

Prevention and control measures

- No measure needs to be taken: the scum layer is totally normal in anaerobic ponds, helping to maintain the absence of oxygen, to control the temperature and to hinder the release of bad odours

Table 10.5. Main operational problems of **facultative ponds** and their possible solutions

---

Problem: scum and floating material (preventing the passage of light energy)

Causes

- Excessive blooming of algae (forming a greenish surface)
- Discharge of unwanted material (e.g.: rubbish)
- Sludge lumps released from the bottom
- Little circulation and wind influence

Prevention and control measures

- Break the scum with water jets or with a rake (broken scum usually sinks)
- Remove the scum with cloth sieves, burying it later
- Break or remove the sludge lumps
- Remove physical obstacles to penetration of the wind (if possible)

Problem: bad odours caused by overload

Causes

- Overload of sewage, causing lowering of the pH, reduced DO concentration, change in the effluent colour from green to yellowish green (predominance of rotifers and crustaceans, which eat algae), appearance of grey zones close to the influent, and bad odours

Prevention and control measures

- Change the operation of the ponds from serial to parallel
- Remove temporarily the problematic pond from operation (provided there are at least two ponds in parallel)
- Recirculate the effluent at a ratio of 1/6
- Consider the adoption of multiple inlets, to avoid preferential paths
- In case of consistent overloads, consider the inclusion of aerators in the pond
- Add occasionally sodium nitrate, as a supplementary source of combined oxygen

Problem: bad odours caused by poor atmospheric conditions

Causes

- Long periods with cloudy weather and low temperature

Prevention and control measures

- Reduce the water depth
- Put a pond in parallel in operation
- Install surface aerators close to the influent inlet

Problem: bad odours caused by toxic substances

Causes

- Toxic substances from industrial discharges, generating sudden anaerobic conditions in the pond

Prevention and control measures

- Perform a complete physical–chemical analysis of the influent, in order to identify the possible toxic compound
- Identify, in the catchment area, the industry causing the discharge, taking the measures provided for by the legislation
- Isolate the affected pond
- Place a second unit in parallel in operation, provided with aeration, if possible

Table 10.5 (*Continued*)

Problem: bad odours caused by hydraulic short circuits

Causes
- Poor distribution of the influent
- Dead zones resulting from the excessive utilisation of the contours when shaping the pond
- Presence of aquatic vegetation in the pond

Prevention and control measures
- Collect samples at several points in the pond (e.g.: DO) to verify whether there are significant differences from one point to another
- In case of multiple inlets, provide a uniform distribution of influent flow in all inlets
- In case of a simple inlet, build new inlets
- Cut and remove aquatic vegetation
- In case of dead zones, introduce aeration to cause small mixing

Problem: bad odours caused by masses of floating algae

Causes
- Excessive blooming of algae, preventing the penetration of light energy, and causing problems associated with the mortality of the excessive population

Prevention and control measures
- Water jet with hose
- Destruction by rake
- Removal by sieves

Problem: high concentrations of algae (SS) in the effluent

Causes
- Environmental conditions that favour the growth of certain algae populations

Prevention and control measures
- Remove the effluent submerged through baffles, which retain the algae
- Use multiple cells in series, with a small detention time in each cell
- Undertake the post-treatment of the effluent from the pond, to remove excessive SS

Problem: presence of cyanobacteria

Causes
- Incomplete treatment
- Overload
- Unbalanced nutrients

Prevention and control measures
- Break the blooming of algae (cyanobacteria)
- Add judiciously copper sulphate

Problem: presence of filamentous algae and moss, which limit
the penetration of light energy

Causes
- Overdesigned ponds
- Influent load seasonally reduced

Prevention and control measures
- Increase the unit load, through the reduction of the number of ponds in operation
- Use operation in series

(Continued)

Table 10.5 (*Continued*)

Problem: progressive tendency to reduce the DO (DO below 3 mg/L in the warm months)

Causes

- Low penetration of sun light
- Low detention time
- High BOD load
- Toxic industrial wastewater

Prevention and control measures

- Remove floating vegetation
- Reduce load in the primary pond through operation in parallel
- Introduce complementary aeration
- Recirculate the final effluent

Problem: progressive tendency to reduce the pH (ideal pH above 8), with mortality of the green algae

Causes

- Overload
- Long periods with adverse atmospheric conditions
- Organisms eating algae

Prevention and control measures

- See measures related to low DO or bad odours due to overload

---

Problem: proliferation of insects

Causes

- Presence of vegetation on the internal slopes of the ponds in contact with the water level

---

Prevention and control measures

- Reduce the water level, causing the larvae trapped in the vegetation of the slopes to disappear when the area dries
- Operate the pond with variation in the water level
- Protect the internal slope with concrete plates, reinforced mortar, geomembrane, etc
- Place fish in the pond, such as carps
- Destroy the scum
- Apply chemical products judiciously

Problem: vegetation inside the pond

Causes

- Low operational level of the pond (below 60 cm)
- Excessive seepage
- Low wastewater flow

Prevention and control measures

- Operate the ponds with a level higher than 90 cm
- Cut the vegetation on the internal borders, preventing it from falling in the ponds
- Protect the slope internally with concrete plates, reinforced mortar, rip-rap, geomembranes, etc
- Remove the vegetation inside the pond with canoes or dredges (lower the water level to facilitate the operation)
- Reduce the permeability of the pond with a layer of clay (if possible)
- Apply herbicides judiciously

Table 10.6. Main operational problems of **aerated ponds** and their possible solutions

Problem: DO absent in some points

Causes
- Poor positioning of the aerators
- Overload in the initial sections

Prevention and control measures
- Change the position of the aerators
- Place more aerators close to the inlet end
- Analyse overloading conditions (see corresponding items in Table 10.5)

Problem: occurrence of bad odours and flies

Causes
- Accumulated scum in the corners and in the internal slopes

Prevention and control measures
- Remove the floating material

Problem: variable DO, dispersed floc and foam

Causes
- Shock loads
- Over aeration
- Industrial wastewater

Prevention and control measures
- Control the operation of the aerators by switching on-off
- Monitor DO to establish the ideal form of operation of the aerators
- Maintain DO around 1 mg/L or more
- Identify the industrial wastewater causing the foams and require its pre-treatment

11

Management of the sludge from stabilisation ponds

11.1 PRELIMINARIES

As in all biological wastewater treatment processes, there is also production of sludge in stabilisation ponds. This sludge is associated with the solids present in the raw sewage and, mainly, with the biomass developed in the biological treatment itself. The various chapters that cover stabilisation pond variants in this book present values for the estimated volumetric sludge production (expressed in m^3/inhab.year or in cm/year). This chapter, based on Gonçalves (1999), presents additional details about the characteristics of the sludge and, mainly, about the management of the sludge from stabilisation ponds. However, the reference Gonçalves (1999) should be consulted for further details on the theme.

One of the main advantages of the *facultative* ponds is the possibility to accumulate sludge on the bottom of the pond, during the whole operational period, with no need for its removal. However, in the most *compact* ponds (anaerobic ponds, facultative aerated lagoons and sedimentation ponds), the occupation of the useful volume of the pond with the accumulated sludge is more significant, requiring an appropriate management, including removal, occasional processing and final disposal.

A further description about the treatment and final disposal of the sludge is not intended here as these items are dealt with in the quoted reference and in Part 7 of this book.

11.2 CHARACTERISTICS AND DISTRIBUTION OF THE SLUDGE IN STABILISATION PONDS

The characteristics of the sludge accumulated in the stabilisation ponds vary according to its retention time in the pond, which usually amounts from some to many years. In this period, the sludge undergoes *thickening* and *anaerobic digestion*, which are reflected on the high contents of total solids (TS) and on the low volatile solids / total solids ratio (VS/TS).

The sludge removed from primary ponds usually presents high contents of total solids, frequently higher than 15%. Because of thickening, the solids concentration varies along the sludge layer, with higher values in the lower parts. Sludges from shallow polishing ponds accumulated over short time periods (one year or less) have average solids concentrations of approximately 4 to 6% (Brito et al, 1999; von Sperling et al, 2002b).

The sludge from ponds operating for several years is usually well digested, with VS/TS ratios lower than 50%.

In terms of nutrients (nitrogen, phosphorus and potassium), the data obtained from an anaerobic pond and from a primary facultative pond (Gonçalves, 1999) suggest nutrient contents lower than those obtained from other wastewater treatment processes. The average values obtained were: TKN: 2.0% of the TS; P: 0.2% of the TS; K: 0.04% of the TS.

With relation to heavy metals the dependence between these characteristics and the presence and type of industrial wastes are also valid here.

Regarding coliforms, the contents in the sludge range between 10^2 and 10^4 FC/gTS, and their decay takes place during the accumulation period in the pond.

Helminth eggs are found in large quantities in pond sludge, since the main egg removal mechanism from the liquid phase is sedimentation. The figures vary substantially from one wastewater treatment plant to another, in view of the variable counting in the raw sewage in each location. Values obtained from the sludge of an anaerobic pond (Gonçalves, 1999) and two polishing ponds (von Sperling et al, 2002), both in Brazil, ranged largely from 30 to 800 eggs/gTS. A long sludge digestion period in the pond seems to contribute to a reduced viability of the eggs. However, it is important to highlight that the sludge from ponds, even after several years, still contains viable eggs, what must be taken into account in their management. Data on the sludge from the anaerobic pond mentioned above, operating for several years, are associated with a percentage of viability between 1 and 10%, while the sludge from the polishing ponds, after operation periods of only six months and one year, presented much higher percentages of viability, between 60 and 90%. The helminth species prevailing in the referred to ponds was *Ascaris lumbricoides*, ranging from 50 to 99% of the total counting of eggs found. On worldwide terms, the most prevailing helminth species is *Ascaris lumbricoides*, but of course the countings and the percentage distribution will vary from place to place.

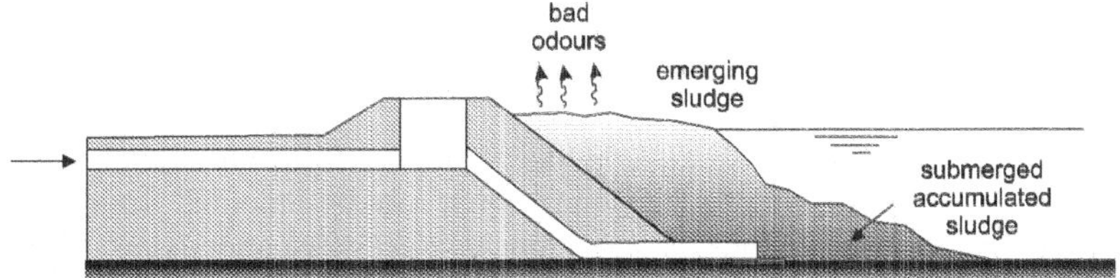

Figure 11.1.   Non-homogeneous spatial distribution of the sludge, with sludge rising to the surface and possible release of malodorous compounds (adapted from Gonçalves, 1999)

The thickness and the characteristics of the sludge layer vary inside the ponds, depending on their geometry and on the positioning of the inlet and outlet structures. Different profile patterns were observed by Gonçalves (1999), but the most frequent one, mainly in primary ponds and in elongated (baffled) ponds, is that of a higher sludge layer close to the inlet. The greatest concern occurs when the sludge layer rises up to and over the water surface, allowing the release of malodorous compounds (Figure 11.1). This situation happens more frequently in ponds without previous grit removal and in anaerobic ponds. In case the sludge is not removed, at least the inclusion of a grit chamber and the redistribution of the emerging sludge layer and of the pond inlets should be performed.

11.3 REMOVAL OF SLUDGE FROM STABILISATION PONDS

11.3.1 Introduction

The removal of sludge is likely to be a compulsory task of a significant scale in the operation of many ponds. However, there is still no widely accepted engineering solution for that. The removal needs to be well planned, since the technique used can change the characteristics of the sludge (increase the water content), and hinder its final disposal.

Gonçalves et al (1999) present in detail the planning and the techniques employed to remove sludge from the ponds. The present item is integrally based on this reference.

11.3.2 Information on the sludge volume to be removed

The planning of the sludge removal from a pond has the purpose of minimising costs, anticipating solutions to occasional problems, and reducing impacts related to the sludge removal and disposal. The following stages are essential in the cleaning operation:

1.   Determination of the pond geometry based on the design or on a topographic survey.

2. Accomplishment of the pond bathymetry, defining bathymetric sections, liquid height of the pond, and depth of the sludge layer.
3. Physical–chemical and microbiological characterisation of the sludge.
4. Definition of the technique to be adopted in the removal of the sludge and, if necessary, in the sludge dewatering and transportation.
5. Definition of the adequate final destination of the sludge, considering the lowest possible environmental impacts.

Certainly, stages 1, 2, and 3 are pre-requisites for the implementation of stage 4, which defines the technique to remove the sludge from the pond. Although there is no consensus on the technique, its selection has a direct impact on the water content of the sludge and, therefore, on the sludge volume to be disposed of later on.

The subsequent items describe the sludge removal stage. The possible sludge processing (dewatering, disinfection) and its disposal are dealt with in Part 7 of this book.

11.3.3 Techniques for sludge removal from ponds

11.3.3.1 Main techniques for sludge removal

The main pond sludge removal techniques can be classified as follows:

- mechanised or non mechanised
- with interruption or no interruption of the pond operation

This second classification was adopted in the following description, due to the importance of the decision of whether to maintain the pond in operation or not.

For the cases in which the sludge should be submitted to dewatering after removal, the following alternatives can be considered: natural drying in the pond itself, use of drying beds, sludge lagoons, or even the use of mechanical equipment.

In locations with a large number of ponds in the surroundings, the use of a mobile dewatering unit (e.g. with centrifuges) could be taken into consideration.

11.3.3.2 Sludge removal with temporary interruption of the pond operation

The temporary deactivation of a pond can be a simple operational measure, if the primary pond stage has been designed in modules, and if there is an idle treatment capacity. However, if this stage consists of a single pond, or if the nominal design load has been already reached, the temporary deactivation may put in risk the stability of the subsequent treatment stage.

Another important aspect is related to emptying the pond. This operation, necessary for drying the sludge in the pond itself, requires previous planning and consent from the environmental agency. In case of very fast emptying, mainly in anaerobic ponds, the impact of the anaerobic effluent on the receiving body can exceed its self-purification capacity. Fish death, unpleasant odours and protests by the population may arise as a consequence.

a)   Manual removal

In this case, the sludge is submitted to drying inside the pond itself, until it is consistent enough to be removed by spades and wheelbarrows (TS>30%).

The disadvantage of this technique is that it requires a long drying period. Considering the period of time necessary to empty the pond, the drying period, and the period for the manual removal of the sludge, the pond will certainly remain deactivated for more than 3 months.

However, the sludge volume to be removed under these conditions is much lower than the volume existent prior to the drying. Another positive aspect is the possible complementary disinfection of the sludge by sunlight-induced pasteurisation. This can be a feasible solution for small sewage treatment plants (<5000 inhabitants).

b)   Mechanical removal (by tractors)

As in the previous technique, the sludge is submitted to drying in the pond and removed soon after. In view of the higher yield of the machines in the sludge removal, the pond can start to work again more quickly than in case of manual removal. However, for tractors or shovels to gain access to the bottom of the pond, the soil support capacity should be previously verified, so that neither the pond bottom sealing nor the stability of the slopes are affected.

The ease of access of the machines into the pond should be evaluated, considering the option of partial rupture of the slopes for further reconstruction. There have been cases of tractors stuck in the sludge in ponds, for which reason it is recommended that the bottom of the pond should not be accessed while the sludge presents a pasty consistency (20% < TS < 30%).

c)   Mechanised scraping and pumping of the sludge

When the pond cannot be deactivated for a very long period of time, the sludge is partially dried in the air, mechanically scraped, and then pumped. This technique requires the aid of a tractor or another device to convey the sludge still in the liquid state to a lower point from where it will be pumped.

The use of positive displacement pumps (piston, diaphragm, rotating lobes, high-pressure piston, etc.) is recommended due to their capacity to move the sludge mass. Torque pumps (centrifuges) can be used, although they require dilution of the highly concentrated sludge, which results in an increased volume of sludge removed.

11.3.3.3 Sludge removal with the pond in operation

a)   Removal by hydraulic sludge discharge pipe

The hydraulic sludge discharge pipe (bottom drain) is the device more frequently included in the design of anaerobic or aerated stabilisation ponds. Nevertheless, it is a solution highly criticised by operators.

There are several reports on clogging and loss of function of this device during the operation of the pond. The problem occurs in view of the evolution of solids contents in the sludge over the years, making its consistency change from liquid to pasty. Should the sludge be discarded with a higher frequency (<5 years), which would prevent its thickening at levels higher than 7% on the bottom of the pond, this device could be useful in small sewage treatment plants. For Victoretti (1975), sludge discharge devices are unnecessary, because the ponds operate for long periods with no need of sludge removal. According to the author, the units should be designed to be deactivated for drainage and removal of the sludge.

In case this technique of pond sludge removal is adopted, pipe diameters equal to or larger than 200 mm are recommended (Metcalf and Eddy, 1991).

b)   Removal by septic tank cleaning truck

Septic tank cleaning or similar trucks are provided with a vacuum suction system with a flexible pipe that removes the sludge and conveys it to the sludge storage compartment in the trucks themselves.

The disadvantage of this solution is that it removes the sludge with a high water level, once pumping requires the dilution of the sludge layers already in an advanced thickening stage. The result can be the need of many trips to transport the sludge from the sewage treatment plant to the disposal site. However, its great advantage is that it removes and transports the sludge in the same operation. The equipment can also be easily found and rented in medium- and large-sized cities.

c)   Dredging

The use of dredges allows the removal of sludge with TS contents higher than 15%, if the sludge is scraped by mechanical means. For sludge with higher solids levels, this type of removal process is affected due to the consistency of the material.

The dredges can also be provided with a sludge-layer-breaking device, so that the removal is accomplished by pumping. In this case, the sludge is removed with water contents higher than those in case of mechanical scraping. Remote control equipment is available.

The dredging may suspend solids at the pond outlet, following revolvement of the bottom sludge layer. This fact can cause a significant load of solids to the secondary pond, if existent. Another important aspect refers to the stability of the waterproofing seal on the bottom of the pond, which may be damaged by the dredging.

Table 11.1. Advantages and disadvantages of the sludge removal techniques from stabilisation ponds

Sludge removal techniques used with deactivation of the pond		
Technology	Advantages	Disadvantages
Manual removal	• Sludge humidity is removed in the pond itself • Cleaning of the pond is done in a controlled way • Sludge with high TS contents reduces transport costs • Almost complete removal of the sludge	• The pond is deactivated for a long period of time • Employees have direct contact with the sludge
Mechanical removal (by tractors)	• Sludge humidity is removed in place • Cleaning of the pond is done in a controlled way • Sludge with high TS contents reduces transport costs • Higher yield than that of manual sludge removal • Almost complete removal of the sludge	• The pond is deactivated for a long period of time • Possible demolition of part of the slope for machine access • The bottom of the pond may be damaged, requiring repairs • Tractor may get stuck in the sludge
Mechanised scraping and pumping	• Shorter sludge drying time in the pond • Almost complete removal of the sludge	• Removal of sludge with a high water content • Requires tractor access in the pond
Sludge removal techniques with the pond in operation		
Technology	Advantages	Disadvantages
Vacuum system from a septic tank cleaning truck	• Operational simplicity • The equipment is easily available • The sludge is removed and transported in the same operation	• Sludge removal with higher frequency – requires low TS contents • Removal of sludge with a high water content due to the mixing with the liquid during the operation • Requires natural or mechanical dewatering of the sludge removed
Hydraulic discharge pipe	• Operational simplicity • Low cost	• Discharge device gets blocked • Sludge discharge with higher frequency – requires low TS contents • Requires natural or mechanical dewatering of the sludge removed • Difficult control of the discharge operation

(Continued)

Table 11.1 (*Continued*)

Sludge removal techniques with the pond in operation		
Technology	Advantages	Disadvantages
Dredging	• Removes the sludge almost completely • Sludge removed with high concentration of solids • Cleaning can be done at a lower frequency	• Need of natural or mechanical dewatering of the sludge removed • Difficult control of the sludge removal operation • Cost of the equipment
Pumping from raft	• Operational simplicity • The equipment is easily available	• Sludge removal with higher frequency – requires low TS contents • Requires natural or mechanical dewatering of the sludge removed • Difficult control of the sludge removal operation
Robotic system	• Removes the sludge almost completely • Sludge with high TS contents reduces transport costs • Allows pond cleaning at lower frequency	• Cost of the equipment • Little availability of the equipment in developing countries

d) Pumping from a raft

The sludge can be pumped from the bottom of the pond by a motor pump installed on a raft. The use of positive displacement pumps (piston, diaphragm, rotating lobes, high-pressure piston, etc.) is also recommended. The motor pump can be propelled by either electricity or fuel. Remote control equipment is available.

The use of centrifugal pumps is only feasible in cases in which the sludge still has a liquid consistency (TS contents <6%), or in cases in which the motor pump is provided with a device for scarifying the sludge on the bottom. The sludge removed by pumping is conveyed outside the pond, where it can be either transported or dewatered in place.

e) Robotic system

This alternative is not largely used in developing countries yet. It can be considered a promising technology in sludge extraction, and consists of a small remote-controlled robotic tractor that moves on a crawler. In the front part of the tractor, the sludge layer is broken and aspired, being then removed from the pond by pumping. The process seems to be capable of removing sludge with high concentrations of solids (TS>20%), allowing the pond to be cleaned at longer intervals. Its main disadvantages are the absence of experience with the equipment in developing countries and the fact that it is imported.

Table 11.2. Comparison of the main factors for selection of the sludge removal technique in stabilisation ponds

Technique	Performance	Ease of operation	Execution time	Sludge volume	Cost
Manual removal	* * * *	* *	* * * *	*	* *
Mechanical removal (by tractors)	* * * *	* * *	* * *	*	* *
Mechanised scraping and pumping	* * *	* *	* * *	* *	* *
Vacuum system from a septic tank cleaning truck	*	* * *	* *	* * * *	* * *
Hydraulic discharge pipe	*	* * * *	* *	* * *	*
Dredging	* * *	* *	* *	* * *	* * *
Pumping from raft	* *	* *	* *	* * *	* *
Robotic system	* * * *	*	* *	* *	* * * *

Scale: * * * * Larger $\rightarrow\rightarrow\rightarrow$ * Smaller

11.3.4 Advantages and disadvantages of the sludge removal techniques

The main advantages and disadvantages of the different sludge removal techniques mentioned previously are summarised in Table 11.1. A comparison among the different techniques considered, involving factors such as process performance, operational ease, flexibility with relation to the final disposal of the sludge, amount of sludge removed, and operational cost, is presented in Table 11.2. The comparison is just for an initial analysis, since the specific conditions of each stabilisation pond can change completely the applicability of the techniques at issue.

References

ABDEL-RAZIK, M.H. (1991). *Dynamic modelling of facultative waste stabilization ponds.* PhD Thesis, Imperial College, 1991.

ABNT (1991). *Projeto de normas para projeto hidráulico-sanitário de lagoas de estabilização.* Minuta do projeto (in Portuguese).

AGUNWAMBA, J.C. et al. (1992). Prediction of the dispersion number in waste stabilization ponds. *Water Research,* **26** (85).

ALEM SOBRINHO, P., RODRIGUES, M.M. (undated). Contribuição ao projeto de sistemas de lagoas aeradas aeróbias para o tratamento de esgotos domésticos. *Revista DAE,* p. 45–62 (in Portuguese).

ARCEIVALA, S.J. (1981). *Wastewater treatment and disposal.* Marcel Dekker, New York. 892 p.

AYRES, R.M., ALABASTER, G.P., MARA, D.D., LEE, D.L. (1992). A design equation for human intestinal nematode egg removal in waste stabilization ponds. *Water Research,* **26** (6), pp. 863–865.

BRITO, M.C.S.O.M., VON SPERLING, M., CHERNICHARO, C.A.L. (1999). Características do lodo acumulado em uma lagoa chicaneada tratando efluentes de um reator UASB. In: *Congresso Brasileiro de Engenharia Sanitária e Ambiental,* **20,** Rio de Janeiro, 10–14 Maio 1999, p. 913–919 (in Portuguese).

BRITO, M.C.S.O.M., CHERNICHARO, C.A.L., VON SPERLING, M. (2000). Relação entre as temperaturas da água e do ar em uma lagoa de maturação na região Sudeste do Brasil. In: *IX SILUBESA – Simpósio Luso-Brasileiro de Engenharia Sanitária e Ambiental,* Porto Seguro – BA, 9 a 14 Abril 2000. Anais eletrônicos (in Portuguese).

CATUNDA, P.F.C., VAN HAANDEL, A.C, LETTINGA, G. (1994). Post treatment of anaerobically treated sewage in waste stabilization ponds. In: *Anaerobic digestion, 7th Symposium,* Capetown, África do Sul, p. 405–415.

CAVALCANTI, P.F.F., VAN HAANDEL, A.C., KATO, M.T., VON SPERLING, M., LUDUVICE, M.L., MONTEGGIA, L.O. (2001). Capítulo 3: Pós-tratamento de efluentes de reatores anaeróbios por lagoas de polimento. In: CHERNICHARO, C.A.L. (coordenador). *Pós-tratamento de efluentes de reatores anaeróbios.* PROSAB/ABES, Rio de Janeiro. p. 105–170 (in Portuguese).

CETESB (1989). *Operação e manutenção de lagoas anaeróbias e facultativas.* Companhia de Tecnologia de Saneamento Ambiental, São Paulo. 91 p (in Portuguese).

COUNCIL OF THE EUROPEAN COMMUNITIES (1991). Council Directive of 21 May 1991 concerning urban waste water treatment (91/271/EEC). *Official Journal of the European Communities*, No. L 135/40.

CRESPO, P.G. (1995). *Tratamento de esgotos: projeto e operação. Os fundamentos.* Apostila. Departamento de Engenharia Sanitária e Ambiental, UFMG (in Portuguese).

CRITES, R., TCHOBANOGLOUS, G. (2000). *Tratamiento de aguas residuales en pequeñas poblaciones.* McGraw-Hill, Colômbia. 776 p.

ECKENFELDER Jr., W.W. (1980). *Principles of water quality management.* CBI, Boston. 717 p.

EPA (1983). *Design manual. Municipal wastewater stabilization ponds.* United States Environmental Protection Agency. 327 p.

GONÇALVES, R. (coord) (1999). *Gerenciamento do lodo de lagoas de estabilização não mecanizadas.* PROSAB – Programa de Pesquisa em Saneamento Básico. ABES, Rio de Janeiro. 80 p (in Portuguese).

GONÇALVES, R.F., NASCIMENTO, C.G., LIMA, M.R.P. (1999). Capítulo 5. Remoção do lodo das lagoas. In: GONÇALVES, R. (coord). *Gerenciamento do lodo de lagoas de estabilização não mecanizadas.* PROSAB – Programa de Pesquisa em Saneamento Básico. ABES, Rio de Janeiro. 80 p (in Portuguese).

GONÇALVES, R.F., SILVA, V.V., TAVEIRA, E.J.A., OLIVEIRA, F.F. (2000). Algae and nutrient removal in anaerobic-facultative pond system with a compact physical-chemical process. In: *I Conferencia Latinoamericana de lagunas de estabilizacion y reuso.* Cali, 24-27 Outubro 2000, pp. 168–176.

JORDÃO, E.P. & PESSÔA, C.A. (1995). *Tratamento de esgotos domésticos.* ABES, 3ª ed. 683 p (in Portuguese).

KELLNER, E., PIRES, E.C. (1998). *Lagoas de estabilização: projeto e operação.* Rio de Janeiro, ABES. 244 p (in Portuguese).

KELLNER, E., PIRES, E.C. (1999). Desenvolvimento de modelo matemático para determinação do perfil vertical de temperatura e do volume útil de lagoas de estabilização. *Revista Engenharia Sanitária e Ambiental,* Vol. 4, no 1, ABES. pp. 84–92 (in Portuguese).

KÖNIG, A.M. (2000). Biologia de las lagunas de estabilisación: algas. In: MENDONÇA, S.R. Sistemas de lagunas de estabilización. McGraw-Hill. Colômbia (in Spanish).

MARA, D.D. (1995). Waste stabilization ponds: effluent quality requirements and implications for process design. In: *3rd IAWQ International Specialist Conference. Waste stabilization ponds: technology and applications.* João Pessoa, PA, 27–31 Março 1995.

MARA, D.D. (1996). Low-cost wastewater treatment in waste stabilisation ponds and waste stabilisation reservoirs in Brazil. In: *Seminário internacional: Tendências no tratamento simplificado de águas residuárias domésticas e industriais.* Belo Horizonte, 6-8 março 1996. DESA-UFMG.

MARA, D.D. (1997). *Design manual for waste stabilisation ponds in India.* Lagoon Technology International Ltd. Leeds.

MARAIS, G.v.R. (1974). Faecal bacteria kinetics in stabilisation ponds. *J. Env. Eng. Div., ASCE,* **100** (EE1), p. 119.

MARA, D.D., ALABASTER, G.P., PEARSON, H.W., MILLS, S.W. (1992). *Waste stabilisation ponds. A design manual for Eastern Africa.* Lagoon Technology International. Leeds. 121 pp.

MENDONÇA, S.R. (1990). *Lagoas de estabilização e aeradas mecanicamente: novos conceitos.* João Pessoa. 385 p (in Portuguese).

MENDONÇA, S.R. (2000). *Sistemas de lagunas de estabilización.* McGraw-Hill. Colômbia. 370 p (in Spanish).

METCALF & EDDY, Inc. (1991). *Wastewater engineering. treatment, disposal, reuse.* 3.ed. McGraw-Hill.

OLIVEIRA, F.F., GONÇALVES, R.F. (1995). Readaptação de ETEs com lagoas de estabilização facultativas a padrões rigorosos de qualidade através de biofiltros aerados submersos. In: *Congresso Brasileiro de Engenharia Sanitária e Ambiental*, 18, Salvador, 17–22 setembro 1995. ABES (in Portuguese).

ORAGUI, J.I., CAWLEY, L., ARRIDGE, H.M., MARA, D.D., PEARSON, H.W., SILVA, S.A. (1995). Pathogen removal kinetics in a tropical experimental waste stabilisation pond in relation to organic loading, retention time and pond geometry. In: 3rd *IAWQ International Specialist Conference. Waste stabilization ponds: technology and applications*. João Pessoa, PA, 27–31 Março 1995.

PANO, A., MIDDLEBROOKS, E.J. (1982). Ammonia nitrogen removal in facultative wastewater stabilisation ponds. *Journal of the Water Pollution Control Federation*, **54** (4), pp. 344–351.

PEARSON, H.W., MARA, D.D., ARRIDGE, H.A. (1995). The influence of pond geometry and configuration on facultative and maturation waste stabilisation pond performance and efficiency. *Wat. Sci. Tech.*, **31** (12), pp. 129–139.

POLPRASERT, C., AGARWALLA, B.K. (1994). A facultative pond model incorporating biofilm activity. *Water Environment Research*, 66 (5), p. 725.

POLPRASERT, C., BHATTARAI, K.K. (1985). Dispersion model for waste stabilization ponds. *J. Env. Eng. Div., ASCE*, **111**, p. 45.

PÖPEL, H.J. (1979). *Aeration and gas transfer*. 2. ed. Delft, Delft University of Technology. 169 p.

SILVA Jr., C., SASSON, S. (1993). *Biologia 2: Seres vivos, estrutura e função*. (Cesar e Sezar). Atual Editora, São Paulo, 2ª ed. 382 p (in Portuguese).

SILVA, M.O.S.A. (1993). *Lagoas de estabilização*. Curso ABES-MG. Belo Horizonte, 22–25 novembro 1993 (in Portuguese).

SILVA, S.A., MARA, D.D. (1979). *Tratamentos biológicos de águas residuárias: lagoas de estabilização*. ABES, Rio de Janeiro. 140 p.

SOARES, C.A.L. (1995). *Curso básico de esgotos. Módulo III. Tratamento*. ABES-MG, março 1995 (in Portuguese).

SOARES, J., SILVA, S.A., OLIVEIRA, R., ARAÚJO, A.L.C., MARA, D.D., PEARSON, H.W. (1995). Ammonia removal in a pilot-scale WSP complex in Northeast Brazil. In: 3rd *IAWQ International Specialist Conference. Waste stabilization ponds: technology and applications*. João Pessoa, PA, 27–31 Março 1995.

THIRUMURTI, D. (1969). Design of waste stabilization ponds. *J. San. Eng. Div., ASCE*, vol. 95, no. AS2.

VAN BUUREN, J.J.L., FRIJNS, J.A.G., LETTINGA, G. (1995). *Wastewater treatment and reuse in developing countries*. Wageningen Agricultural University.

VAN HAANDEL, A.C., LETTINGA, G. (1994). *Tratamento anaeróbio de esgotos. Um manual para regiões de clima quente* (in Portuguese).

VICTORETTI, B.A. (1975). Capítulo 11: Manutenção e operação de lagoas de estabilização. In: *Lagoas de estabilização*. 2ª ed. CETESB, São Paulo (in Portuguese).

VIDAL, W.L. (1983). Aperfeiçoamentos hidráulicos no projeto de lagoas de estabilização, visando redução da área de tratamento: uma aplicação prática. In: *Congresso Brasileiro de Engenharia Sanitária e Ambiental*, 12, Camboriú, 20–25 novembro 1983. ABES (in Portuguese).

VON SPERLING, M. (1983). *Autodepuração dos cursos d'água*. Dissertação de mestrado, DES- EEUFMG. 366 p (in Portuguese).

VON SPERLING, M. (1995). Design of facultative ponds based on the uncertainty analysis. *Water Science and Technology*, **33** (7). pp. 41–47.

VON SPERLING, M. (1996). *Introdução à qualidade das águas e ao tratamento de esgotos*. 2ª ed. DESA-UFMG, 243 p (in Portuguese).

VON SPERLING, M. (1999). Performance evaluation and mathematical modelling of coliform die-off in tropical and subtropical waste stabilization ponds. *Water Research*, **33** (6). pp. 1435–1448.

VON SPERLING, M. (2001). Remoção de DBO em 12 lagoas de estabilização primárias e secundárias no Sudeste do Brasil. In: IX Simpósio Luso-Brasileiro de Engenharia Sanitária e Ambiental – SILUBESA, Anais eletrônicos, Porto Seguro, abril 2001 (in Portuguese).

VON SPERLING, M. (2002). Relationship between first-order decay coefficients in ponds, according to plug flow, CSTR and dispersed flow regimens. *Water Science and Technology*, **45** (1). pp. 17–24.

VON SPERLING, M. (2003). Influence of the dispersion number on the estimation of faecal coliform removal in ponds. *Water Science and Technology*, **48** (2). pp. 181–188.

VON SPERLING, M., CHERNICHARO, C.A.L., SOARES, A.M.E., ZERBINI, A.M. (2002). Coliform and helminth eggs removal in a combined UASB reactor – baffled pond system in Brazil: performance evaluation and mathematical modelling. *Water Science and Technology*, **45** (10). pp. 237–242.

VON SPERLING, M., CHERNICHARO, C.A.L., SOARES, A.M.E , ZERBINI, A.M. (2003a). Evaluation and modelling of helminth eggs removal in baffled and unbaffled ponds treating anaerobic effluent. *Water Science and Technology*, **48** (2). pp. 113–120.

VON SPERLING, M., JORDÃO, E.P., KATO, M.T., ALEM SOBRINHO, P., BASTOS, R.K.X., PIVELLI, R.P. (2003b). Capítulo 8: Lagoas de estabilização. In: GONÇALVES, R.F. (coordenador). *Desinfecção de esgotos sanitários*. PROSAB, Rio de Janeiro (in Portuguese).

WEF (1990). *Operation of municipal wastewater treatment plants*. MOP 11, Vol. 2. Water Environment Federation.

WHO (1989). *Health guidelines for the use of wastewater in agriculture and aquaculture*. Technical Report Series 778. Geneva: World Health Organization.

WPCF (1990). *Natural systems for wastewater treatment. Manual of Practice FD-16*. Water Pollution Control Federation. Alexandria. 270 p.

YANEZ, F. (1993). *Lagunas de estabilizacion. Teoria, diseño y mantenimiento*. ETAPA, Cuenca, Equador, 421 p (in Spanish).

IWA Publishing's authorised EU representative for General Product
Safety Regulations is Diane D'Arras, 15 rue Duret, 75116 Paris,
France, e-mail: safety@iwap.co.uk.

Printed and bound by CPI Group (UK) Ltd, Croydon, CR0 4YY

12/05/2026

02108891-0008